浓情地中海

Mediterranean

何心 译

Edited by HI-DESIGN PUBLISHING
Dalian University of Technology Press

图书在版编目 (CIP) 数据

浓情地中海：英汉对照 / 高迪国际出版有限公司编；何心译. —大连：大连理工大学出版社，2013.2
ISBN 978-7-5611-7467-8

Ⅰ. ①浓… Ⅱ. ①高… ②何… Ⅲ. ①住宅－室内装饰设计－世界－图集 Ⅳ. ① TU241-64

中国版本图书馆 CIP 数据核字 (2012) 第 288460 号

出版发行：大连理工大学出版社
（地址：大连市软件园路 80 号 邮编：116023）
印　　刷：利丰雅高印刷（深圳）有限公司
幅面尺寸：285mm × 380mm
印　　张：42
插　　页：4
出版时间：2013 年 2 月第 1 版
印刷时间：2013 年 2 月第 1 次印刷
策划编辑：袁　斌 刘　蓉
责任编辑：刘　蓉
责任校对：王丹丹
封面设计：梁文静

ISBN 978-7-5611-7467-8

电话：0411-84708842
传真：0411-84701466
邮购：0411-84703636
E-mail:designbooks_dutp@yahoo.cn
URL:http://www.dutp.cn

如有质量问题请联系出版中心：（0411）84709246 84709043

销售商：上海筑邦文化传播有限公司

Mediterranean

PREFACE

序 言 1

FABRIZIA FREZZA
法布雷西亚・弗雷扎

My architectural school of thought consists in maintaining the original characteristics of the design before renovation and in new constructions. My expression embodies the spirit of the community, its culture, its location including landscape, the psychological human element and studying each area. The "genius loci" is felt in all my works. I blend modern and classical style.

A 20th century palazzo is reborn with a modern, contemporary mix of interior design. I unite each architectural style, for example Mediterranean arches and vaults. I work with artisans from the area and local materials to generate a personalized and integrated project keeping the traditions of the location alive, strongly linking the past with the present.

My projects begin with a thorough study of space and conclude with the rational division of the space to satisfy the needs of my clients. At the base of all my projects where energy flows freely without obstacles is the rationalization of space planning.

My goal is to create an environment where my clients feel at home in the hotels I design and where guests feel comfortable in the homes that I create. I try to bring about harmony through the use of shapes, fabrics and colors in each room. The subtle use of uniform tones from the flooring to the walls generates a neutral venue where the furnishings and accessories are set off by the chromatic uniformity. I create hideaways and relaxing environments never forgetting the practical domestic use of space.

Each project is unique and dealt with individually. Projects are completed from A-Z, from the original plan to the renovation, to the interior design and to the landscape architecture. Each project becomes a journey that begins with a vision and ends with a realization.

在我的建筑作品中，对于旧建筑改建规划的理念是尽量维护和保持原有建筑的设计风格，并在改造后的建筑结构中得以充分体现。这里所说的风格包括周边社区的理念和风俗、建筑的文化底蕴、周边的风景、人文心理以及对于每个细节的重视。在每个作品中都尽量做到现代风格与经典风格的融合，充分体现个人的设计天赋。

20 世纪宫殿式建筑风格重新焕发青春，内饰多为现代与当代风格相互混搭的效果。我在设计中尽量尝试将各种建筑风格进行混合搭配，如地中海拱形门和穹顶等。同时，与建筑工匠合作，充分利用具有当地风格特色的建筑材料，建造富有特色、贯通古今，并能表现当地生活传统的建筑。

对于一个建筑工程来说，首要的是对建筑空间进行全面详尽的研究，尤其是如何根据客户的需求进行空间上的分配。在我的设计方案中，所有的建筑空间设计都尽量趋近最合理的布局，通行顺畅，毫无阻滞。

顾客在我设计的旅馆中感到宾至如归，主人在我设计的住宅中感到身心舒适，这就是我对于建筑设计的最终目标。尽量让每一个空间的形状、质地和颜色达到和谐统一的效果，地板和墙体使用相似风格的色调，打造一种相对中性的背景，从而利用色差来突出色彩鲜明的家具或其他配饰。在突出空间的私密性和休闲娱乐的同时，保证了居家使用空间的比例。

每一个作品都是独一无二的。从最初的规划、逐步改造、内饰设计、直至最后的景观规划，每个项目都是一次完整的历程，始于各种设计创想，以建筑施工完成而结束。

PREFACE

序 言 2

TOMMY CHAMBERS
汤米·钱伯斯

The term "Mediterranean Style" seems to conjure up in the minds of many people an impression of easy living. Whether it's the simple more primitive elements, or the exotic and more opulent ones, the Mediterranean style is easy to adopt.

It is a style that lends itself to a wide variety of new and old building types. A style that is represented by a collection of elements derived from the vernacular architecture of Italy, Spain, North Africa, Morocco, the South of France, and other Mediterranean countries.

What is it that peaks our interest and affection for the Mediterranean experience?

Is it our attraction to the fantasy and poetry of the Mediterranean lifestyle, or the sentimental romantic notion that we see playing out in the charming villages on the shores and steep slopes of the Mediterranean?

Is it the playful and grounded use of architectural design elements such as the verandas, courtyards, tile roofs, domes, the gardens, niches, loggias, terraces, portals and fountains that are so inherent to the modern and ancient Mediterranean way of life?

Or perhaps it's the desire to escape to a time that was simpler and connected to the earth and sea below us, a time that was anchored in nature and to those architectural elements that arose from need and necessity of a culture.

These are all good reasons to love and appreciate the Mediterranean style. And there are so many more. Whatever your reasons are, enjoy exploring the spirit and atmosphere of the Mediterranean.

Tommy Chambers

Tommy Chambers Interiors, Inc.

在很多人的心目中，所谓地中海风格是舒适生活的一种象征。无论是简单至原始时代的设计元素，还是更为奢华的舶来品，地中海风格始终体现的是一种简单舒适的生活方式。

地中海风格适用于广泛的、不同形式的建筑体系，源自意大利、西班牙、北非、摩洛哥、法国南部以及其他地中海国家，代表了一种融合了各种浓郁的乡土气息的建筑类型。

是什么原因促使我们对地中海风格如此着迷呢？

是因为地中海生活方式的美丽幻想或者诗情画意深深吸引了我们？还是那种渗透在地中海海滨或是悬崖峭壁上的迷人乡村中淡淡伤感的浪漫主义情怀深深感染了我们？

是走廊、庭院、瓦片拱形屋顶、花园、壁龛、凉廊、阳台、桥门、喷泉，这些或休闲或坚固的建筑设计元素与生俱来地体现了现代或古代地中海地区的生活方式？

也许是我们内心深处渴望远离喧嚣、去亲近土地和大海的欲望；或是源自文化底蕴中驻足大自然、潜移默化的天人合一的境界。

这些都是我们喜爱并欣赏地中海风格的原因所在，其实原因还有很多很多。无论如何，希望您能不断探求这种地中海式生活方式的精神和氛围，充分享受这一美好的过程。

汤米·钱伯斯

汤米·钱伯斯室内设计公司

CONTENTS

目 录

PARROT TREE

Casa Paralea

帕梅拉住宅

LOCATION | St. Lucie River in Stuart, Florida, USA

Old World charm and grace beautifully describes this magnificent home designed in the classic Mediterranean style of Addison Mizner.

The design of the home was inspired by the owners' trip to Italy. During their tour of Tuscany they fell in love with the Tuscan region and the architecture. Upon returning to the United States, they decided to build their dream home.

The house features over 929 square meters of living space on two floors with a third floor tower to view the Atlantic Ocean. Venetian plaster, marble floors and custom hand painted murals are throughout the main floor and master retreat.

Every detail for this home was carefully planned and executed. Custom tiles and marble were made in Italy for the master bathroom. Exotic marble and granite were also used for all of the bathrooms and the summer kitchen. The exquisite bronze lion statues, which greet visitors at the porte cochere, were hand made in Italy. This home is truly a feast for the eyes.

CASA
PARALEA

阿迪森·麦兹那设计的这栋高雅住宅秉承了迷人而优雅的古代建筑艺术气息，具有典型的地中海古典风格。

这栋住宅的设计灵感来自于房主在意大利托斯卡纳地区的短途旅行经历，那段时间，他们完全被托斯卡纳地区的美丽风景和古老建筑所折服。回到美国后，他们决定按照这样的风格来设计建造他们梦中的家园。

整栋住宅下面两层中，可居住面积超过929平方米，第三层设计为瞭望塔的模式，在那里可以直接欣赏大西洋的美丽风景。一层和休息室使用了大量的威尼斯灰墙、大理石地板以及当地手工绘制的壁画作为装饰。

设计师们对住宅的每一个细节进行了精心的设计和施工。主浴室使用了具有意大利装饰特色的传统瓷砖和大理石。颇具异国情调的大理石和花岗岩装饰贯穿于整个浴室空间和户外厨房。大门入口处矗立着一尊精致的青铜狮子雕塑，源自意大利，全手工制作。总体而言，整栋住宅设计独特，堪称视觉上的饕餮盛宴。

Rustic Mediterranean Beach House

地中海乡村海滩别墅

INTERIOR DESIGNER | *Tommy Chambers*
DESIGN COMPANY | *Tommy Chambers Interiors, Inc.*
ARCHITECTURE | *Arc Design Group, Inc.*
CONSTRUCTION | *MS Nolan and Associates, Inc.*
LOCATION | *Hollywood Beach, Ventura County, California, USA*
AREA | *465 m²*
PHOTOGRAPHER | *David Phelps*

This Mediterranean escape for a busy Beverly Hills family makes for the perfect get away. It is all new construction built on a challenging narrow lot on Hollywood beach in Ventura County. Designers were able to create eclectic, comfortable and family friendly interiors while sticking to a pre-determined second home budget.

ART HISTORY
N.C. WYETH
FROM IMPRESSIONISM TO MODERN ART
ROME
ART & ARCHITECTURE
PICASSO
ISAMU NOGUCHI
THE PORTRAITS OF DUANE MICHALS 1958 1988

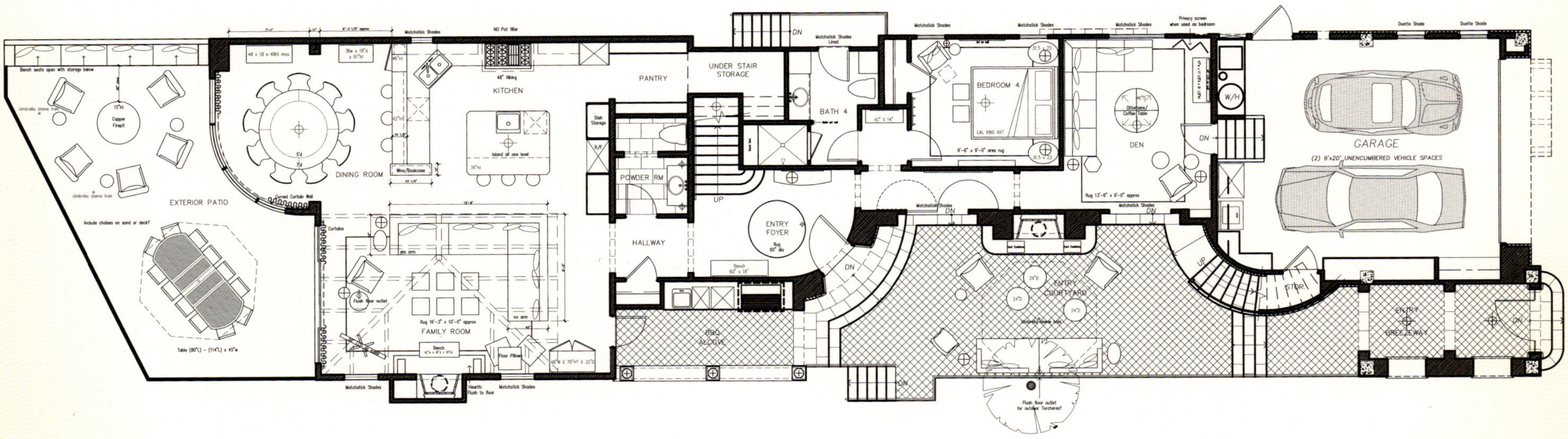

FIRST FLOOR PLAN

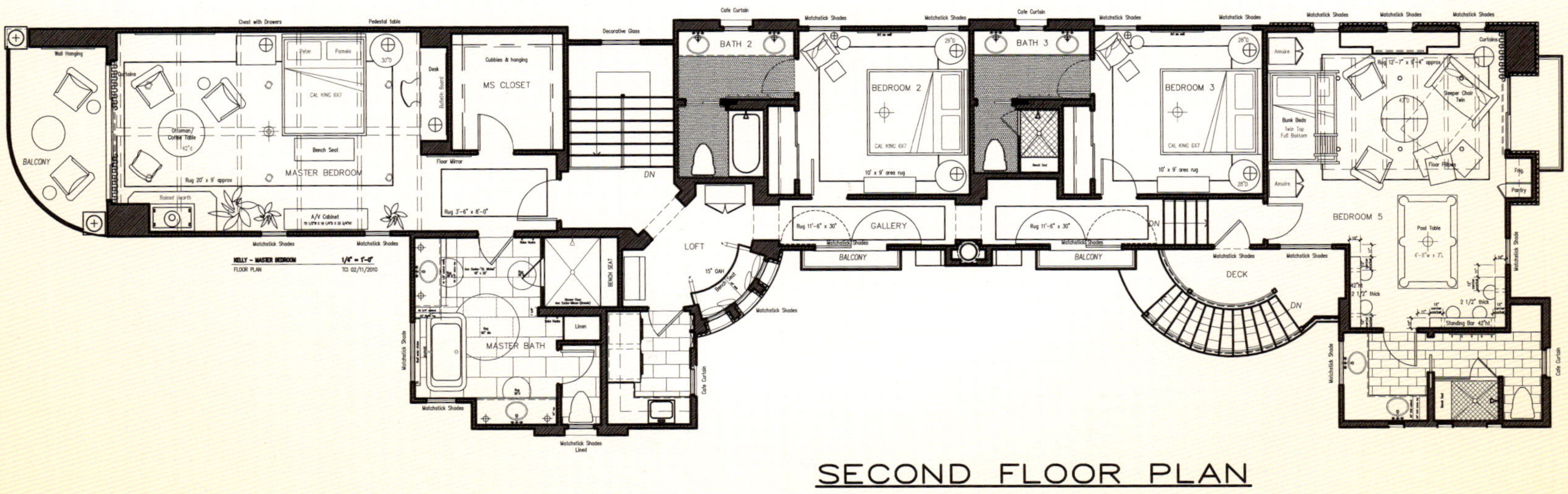

SECOND FLOOR PLAN

为了逃避忙碌的贝佛利山庄的生活，怀揣着完美生活的梦想，一家人举家来到凡图拉郡的好莱坞海滩，在这片狭小的海滩上兴建了这座地中海式建筑。当您决定在这里预定并建设一个新的家园时，设计师们将竭尽全力创造出折中的、舒适的、充满家庭友好氛围的内部环境。

PLATT LYNES

Solcito

索尔希托住宅

ARCHITECT | Jon Poetzl
FIRM | Carson Poetzl, Inc.
LOCATION | Paradise Valley, Arizona, USA
AREA | 4,431 m^2
MATERIAL | Wood Stud, Stucco, Hand Forged Iron, Tile, Clay Roof Tile
PHOTOGRAPHER | High Res Media

This Tuscan retreat was designed for the family that entertains. The designer finishes the house including mixtures of natural stone, canterra, granite, and alder wood which come together to create a timeless design. The open courtyard entry with attached Guest suite invite you inside where one experiences an indoor/outdoor living experience with a 4.88 meters pocket door leading out to an expansive covered patio creating an invisible line between the interior of the home and the resort style backyard. The outdoor living experience features a private pool, spa, water features, outdoor bars, refrigerator, and an observation deck with 360 views of Paradise Valley.

这处幽静的托斯卡纳式建筑是为家庭娱乐而专门设计建造的。设计师采用天然石料、花岗岩和赤杨木等多种建材混合的方式，试图创造一种永恒的设计。开放式的庭院入口处设有客人套房，让人体验一种室内与室外交错的生活感受。踏过 4.88 米的折叠门，便进入一个宽敞的覆盖露台，这样的设计巧妙地将室内居住区与室外休闲娱乐功能的后院分隔开，让人顿感豁然开朗。特色的室外活动体验包括私人游泳池、矿泉疗养、水系景观、室外酒吧、冷藏室和 360 度俯瞰伊甸园峡谷全景的观景台。

Rough Hollow Showcase Home

简约的中空展厅式住宅

ARCHITECT | *Bob Wetmore, AIA*
FIRM | *Cornerstone Architects*
INTERIOR DESIGNER | *David Bravo*
LOCATION | *Austin, Texas, USA*
AREA | *630 m^2*
BUILDER | *Terry Polston, New Austin Homes*
PHOTOGRAPHER | *Coles Hairston*

Winner of five awards in the Rough Hollow Parade of Homes, this 630 square meters home is an exquisite addition to the prestigious Lakeway neighborhood. The Santa Barbara style home features a welcoming colonnade, lush courtyard, beautiful casita, spacious master suite with a private outdoor covered terrace, and a unique Koi pond beginning underneath the wine room glass floor and continuing to the outdoor living area. In addition, the views of Lake Travis are unmatched throughout the home.

FRONT ELEVATION
#06140

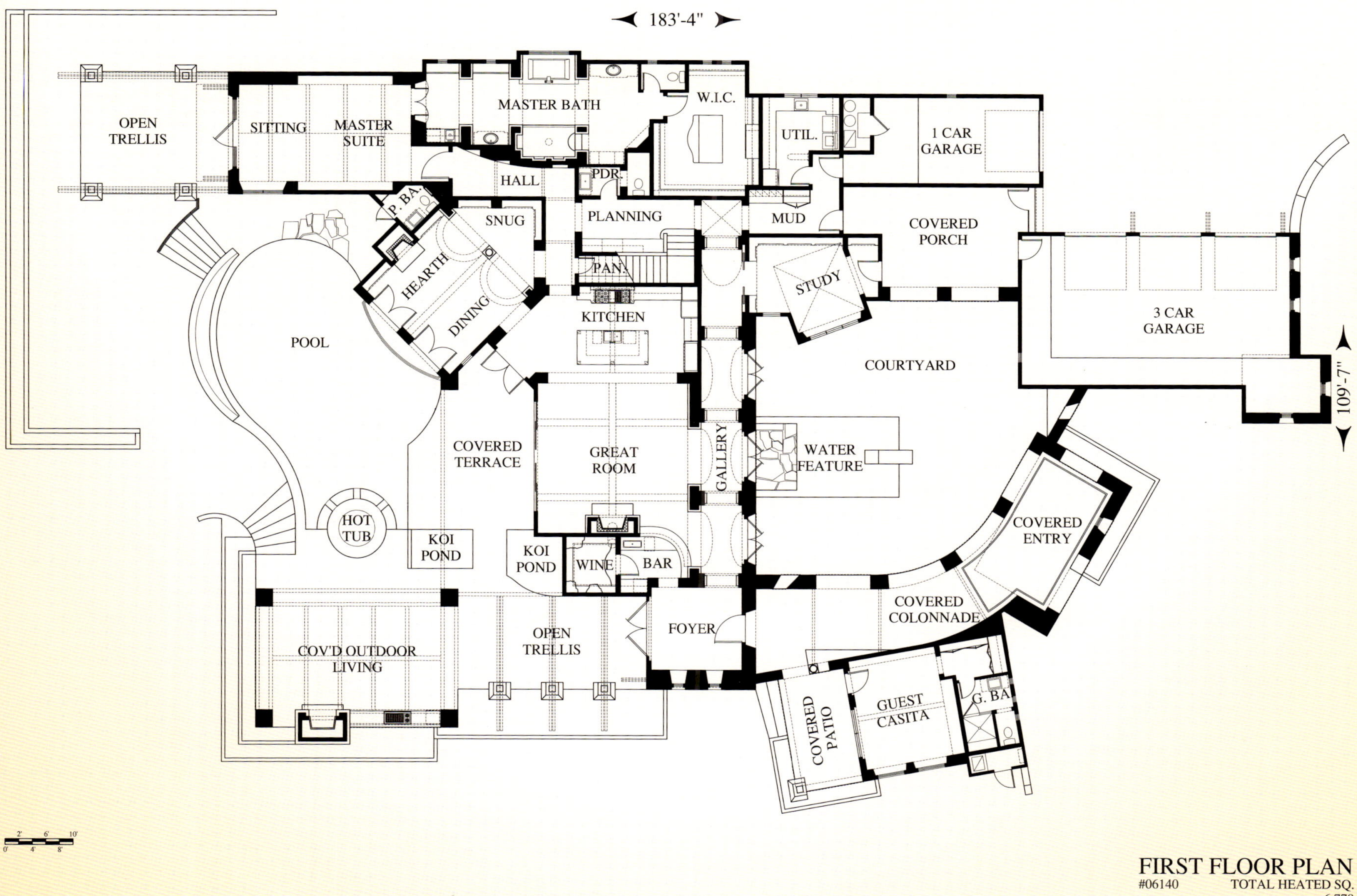

FIRST FLOOR PLAN
#06140 TOTAL HEATED SQ. 6,778

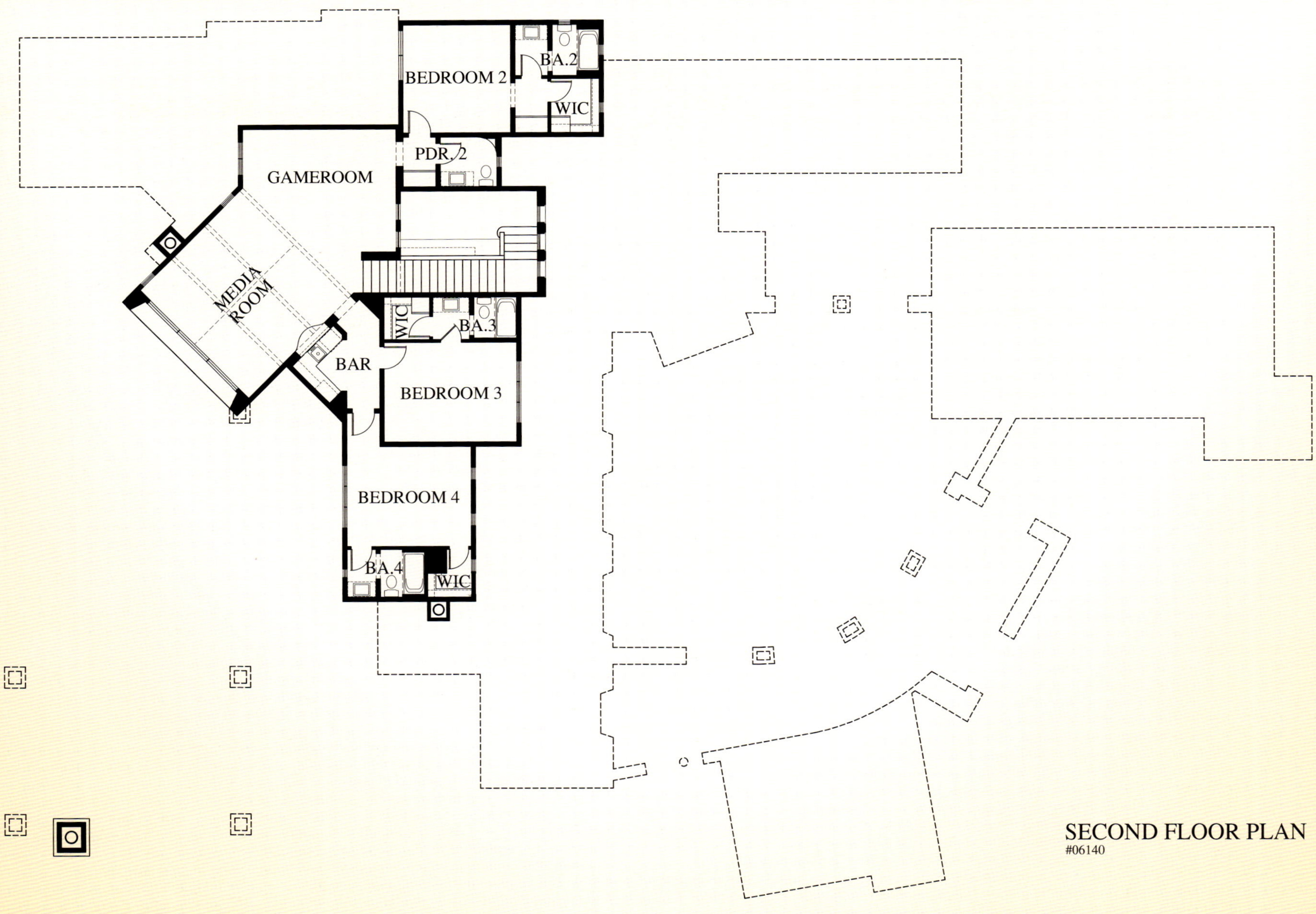

SECOND FLOOR PLAN
#06140

本设计在“简约式中空展厅住宅项目”中荣获五项大奖。住宅占地630平方米，以其优美的设计在名声显赫的莱克威社区中闻名遐迩。圣巴巴拉风格的住宅设有迎宾门廊、郁郁葱葱的庭院和美丽的小屋，宽敞的主人套房带有私人户外露台，独特的锦鲤池塘上面是铺有透明玻璃地板的酒窖，玻璃地板一直延续到户外的活动区域。此外，在住宅内部的很多地方都可以观赏到附近特拉维斯湖令人窒息的美丽风光。

Del Mar Meadows

德玛尔牧场社区住宅

ARCHITECT | Kurt Erlbeck
FIRM | Elevation Architectural Studios
LOCATION | Aceituno St. San Diego, USA
AREA | 557 m²
PHOTOGRAPHER | Martin Mann

Designed to take advantage of it's corner lot and slightly elevated site, this estate is one of the most luxuriously appointed homes in the prestigious Del Mar Meadows community. Taking its design cues from the hillside villages that are artistically painted within the landscape; this collage of different building masses and volumes with a combination of stucco and stone veneer makes the home appear to have been built over generations. A tastefully elegant adaptation of Italy's romantic rural architecture makes this home the jewel of the Meadows Del Mar.

A custom wrought iron gated entry courtyard with a Cantera stone fountain and Italianate cast stone balustrades creates an elevated entry patio that accentuates the powerful front door of this medieval inspired castle tower.

A double-arched stained, distressed wood door creates a rustic, yet formal entry experience to this estate. Entering the rotunda, graceful stone stairs highlights the space, with custom tiled stair risers, artfully designed curved wrought iron handrail and a hand blown glass fixture shipped back from the clients shopping spree in Italy.

A private courtyard with a large outdoor fireplace, hand carved Cantera stone fountain and wooden trellis gate creates a private sanctuary for daily escapes from the outside world as well as creates a spectacular gathering space for larger public events. The living room, library, dining room and family room all open up the courtyard allowing for a wonderful flow from the inside of the home to the outside of the home.

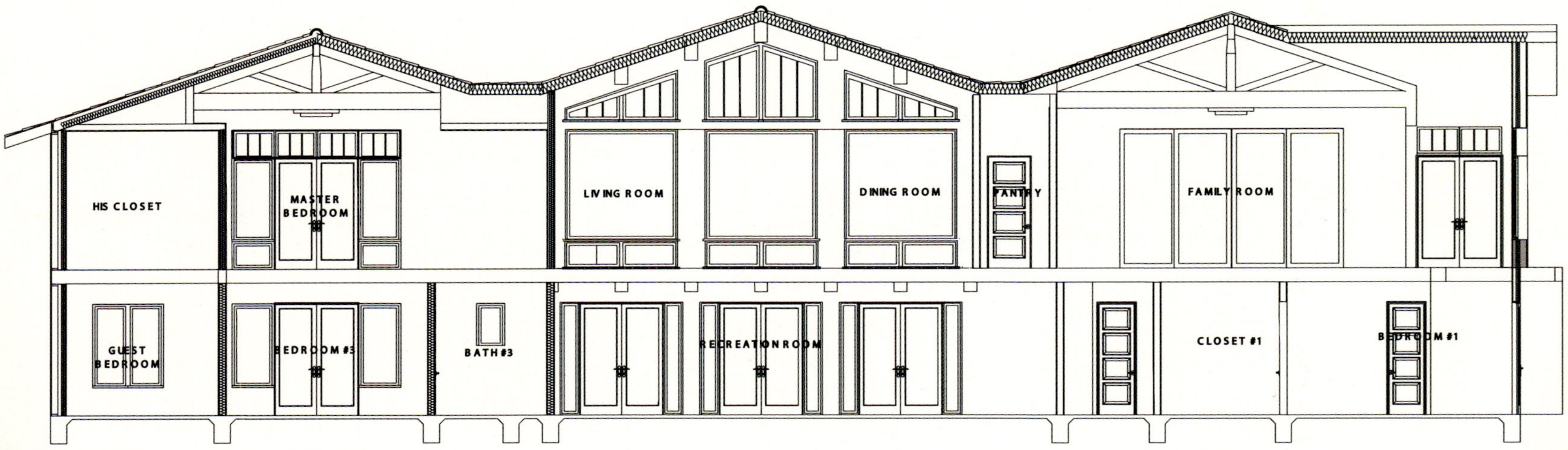

SECTION 1/16" = 1'-0"

EAST ELEVATION 1/16" = 1'-0"

SOUTH ELEVATION 1/16" = 1'-0"

WEST ELEVATION 1/16" = 1'-0"

NORTH ELEVATION 1/16" = 1'-0"

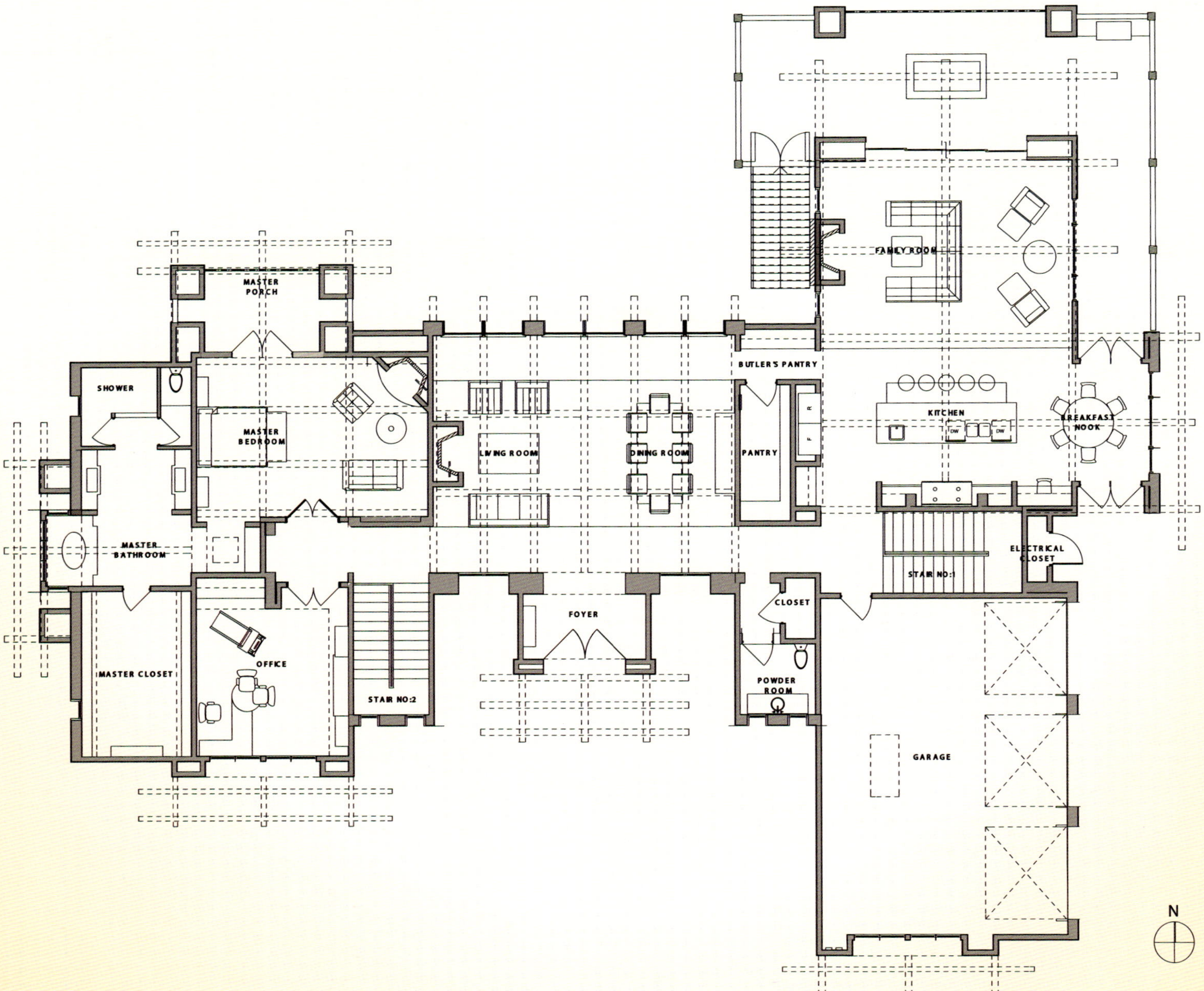

Main Level 1/16" = 1'-0"

这栋建筑位于极富盛名的德玛尔牧场社区中地理位置绝佳的街角高地上，是本社区最为奢华的住宅之一。住宅的设计风格受山区小镇的影响颇深，外墙壁上以艺术的手法描绘了该地区美丽的风景。在建筑外墙上大量采用粉饰灰泥和石质饰板形成的抽象拼贴画，有一种古色古香的味道。整个建筑看起来给人一种饱经风霜、历尽岁月沧桑的感觉。这栋建筑的装饰充满了高雅优美的意大利乡村建筑的浪漫风情，在整个德尔玛牧场社区独树一帜。

住宅的大门是用传统的熟铁工艺制成，庭院内设有一个坎特式石质喷泉。另外，利用意大利风格的铸石砌成的栏杆围绕出一个高出地面的天井入口，分外强调了这座中世纪风格的城堡塔楼正面的威严气势。建筑的木质大门呈双拱形，古老而斑驳，是典型的乡村建筑入口的装饰造型。大门里面是宽敞的圆形大厅，旁边是造型优雅的石质楼梯，造型古典，熟铁栏杆扶手呈弧形，侧面的玻璃衬板显然是借鉴了典型的意大利商用楼梯的构造。

私密的庭院内有巨大的室外壁炉、手工雕刻的坎特式石质喷泉和实木格架式的大门，所有这些建筑元素无不体现了一种远离喧嚣的外部世界，营造温馨的私密空间的理念，同时也为业主提供了一个富丽堂皇的空间，用以举办大型的公众活动。起居室、图书室、餐厅和家庭活动室都朝向庭院，既保持了室内外空气的流通，也极大地增强了室内和室外的互动性。

Kalithea

卡利西亚住宅

INTERIOR DESIGNER | Deborah Costa
DESIGN COMPANY | StageCrafters
ARCHITECT | Gary Randall Shipman
LOCATION | Sacramento, CA, USA
AREA | 604 m^2
PHOTOGRAPHER | Trevor Jobson

The floor plan & outdoor living spaces of this exceptional Mediterranean home flow seamlessly together integrating the serenity of the landscape into the interior. In addition to 5 bedrooms and 5 full bathrooms, this 588 square meters home features a gourmet kitchen, a huge game room, two laundry rooms, office, and extensive tile and stonework. Outside living spaces include a covered loggia off the master bedroom, a lower covered loggia with fireplace and outdoor kitchen off the main living area, upstairs vista off the game room and side courtyard with access from the dining room.

建筑楼层的设计和户外的生活空间充分显示了浓郁的地中海风情。安详宁静的乡村风景与室内设计浑然天成地交融在一起。除了五间卧室和五间独立的浴室外，这里还有588平方米的休闲空间，使用瓷砖进行大范围的装饰，包括一个美食厨房、巨大的游戏室、两间洗衣房、办公室以及石雕作品。建筑外部的活动空间包括一个主卧室外带有顶棚的凉廊，一个高度稍低些通往主要活动区域并设有壁炉和室外厨房的凉廊，以及一条紧邻庭院并连接楼上游戏室和餐厅入口的楼梯。

Lampade
MOSSANT

OLIVER!

1920
MIS EN BOUTEILLE AU CHATEAU
CHATEAU MONT-REDON
Chateauneuf du-Pape

Parrot Tree Plantation

鹦鹉树种植园

ARCHITECT | *Cathi House*
FIRM | *House + House Architects*
LOCATION | *Roatan, Honduras*
AREA | *679,872 m^2*
PHOTOGRAPHER | *Steven & Cathi House*

Parrot Tree Plantation is a unique mixed-use community occupying a 679,872 square meters site of steep hillsides, 4,828 meters of shoreline, sandy beaches and two sheltered bays, on the reef-laced waters of the Caribbean Sea. The master plan layout is influenced by villages on the Greek Islands, the Italian and Dalmatian coasts and hill towns in central Italy and Spain. Commercial spaces and housing are woven into gardens, plazas and winding roads and walkways. The waterfront promenade, lined with shops, offices, cafe, housing and pedestrian gathering spaces, is an active, vibrant focal point, embracing a 200-slip, full-service marina.

Condominium units include myriad plans for richness and variety in the buildings. Arrangements of the individual units and terraces are articulated to respond to the sun and breezes and to respect views and privacy. Pedestrian movement is filled with twists and turns, infusing surprises in every step, offering places that invite social interaction and foster a sense of community. A murky lagoon has been transformed into a luminous 20,234-meter turquoise salt water swimming area surrounded by palm-dotted white powder sand. Private homes line the sandy beaches and climb the hillside above. Beachfront bungalows offer luxurious accommodations that supplement the planned 150-room hotel. An office-reception facility with parking and meeting rooms, a conference center, restaurants, yacht club, sports facilities, health and spa center and hiking trails through this tropical paradise encompass what has been hailed as one of the most innovative developments in Honduras, receiving the prestigious Copan Award.

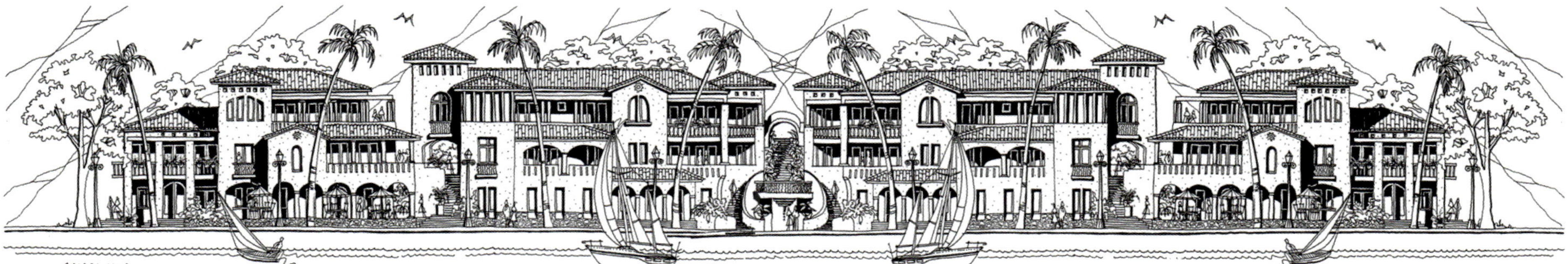

鹦鹉树种植园坐落在满是暗礁的加勒比海岸，是一处设计独特的多功能社区。种植园建在陡峭的山坡上，占地面积 679872 平方米，占据了 4828 米的沙滩海岸线，两处避风港湾。种植园的主体设计方案借鉴了多处建筑元素的特点，比如希腊岛屿上的村庄、意大利人和达尔马提亚人的海岸建筑、意大利和西班牙中部地区山坡上的城镇建筑等。商业区和住宅区巧妙地穿插于园林、露天广场、蜿蜒的街道和小路中间。海滨步道沿线两侧是商铺、办公室、咖啡厅、住宅和供行人集会的场所，一派生机勃勃的景象。这里还有一个可容纳 200 艘游艇并可提供全方位服务的小船坞。

公寓中的多元化设计适合居住于此的富人和各种身份的人群。独立单元和联排单元通过合理的布局形成一个整体，保证了良好的采光和通风，提供了宽阔的视野的同时也保持了住宅的私密性。人们可以在蜿蜒曲折的小路上游逛，感受着处处可见的惊喜。种植园内还提供举办社交活动的场所，在那里可以增进人们的社交观念。原本暗淡无光的咸水湖已经被改造成占地面积约 20234 平方米，由棕榈树点缀，白色沙滩环抱的清新明亮、绿波荡漾的海水游泳浴场。私人住所沿着沙滩一线排开，一直延伸到山坡上。海滩上的单层小屋提供了奢华的住宿设施，可以作为拥有 150 间客房的酒店的有益补充。这里还可以为办公接待提供完备的服务，如停车场、会议室、会议中心、饭店、游艇俱乐部、运动设施、健康和矿泉疗养中心等等。徒步穿越热带伊甸园——这项服务被誉为洪都拉斯最具创新性的发展之一，并因此被授予了科潘奖。

5700 Spanish Oaks

5700 棵西班牙橡树中的住宅

DESIGNER | *John Hathaway, AIA*

DESIGN COMPANY | *Vanguard Studio Inc.*

LOCATION | *Austin, TX, USA*

AREA | *409 m^2*

PHOTOGRAPHER | *James Marzano*

This custom built one-story home is in the heart of the exclusive community of Spanish Oaks. The centerpiece of this home is the beautiful courtyard, offering views to its centrally located pool from nearly all rooms in the house. This courtyard in the old world tradition brings new life to the concept of outdoor living. The central idea for this house is hospitality. The house features a large outdoor area with a unique covered breezeway that provides the main source of entertainment. Continuing on the theme of entertainment, the home showcases a combined living room, dining room, and kitchen. This provides a larger open room where people can eat, watch television, and talk. The kitchen is equipped with state-of-the-art appliances to host guests and to feed a family. Also giant sliding doors in the living room make the inside flow right into the outside. When you are not entertaining, the pool provides a relaxing environment with golf course views. The Spanish Oaks home takes great pride in its internal focus; throughout the house, classical stone detail is at its finest. Builders Art Duncan and Scott Bryant of Emerald Crest brought their skills to bear in finishing this beautiful home.

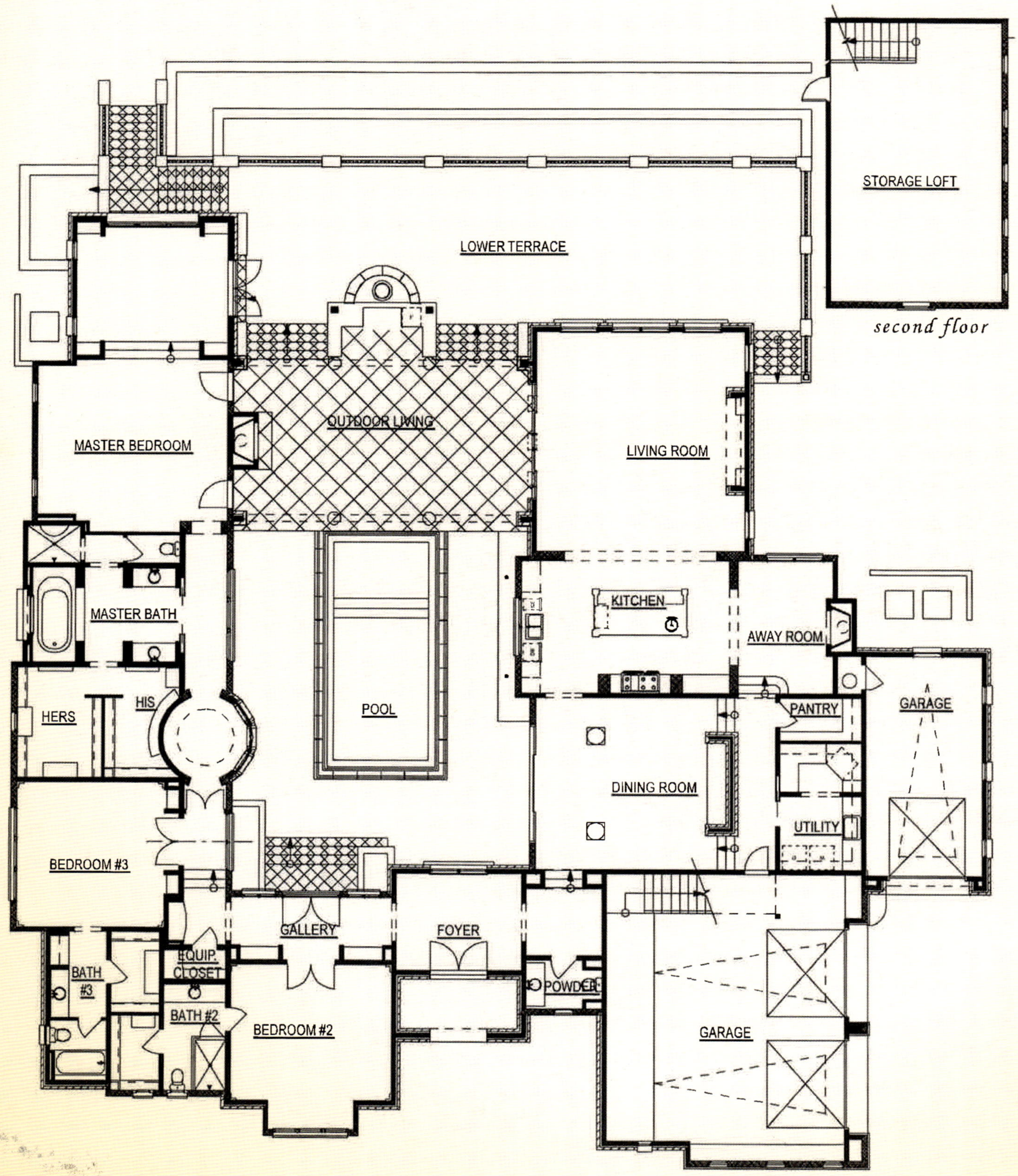

second floor

first floor

这栋单层古典住宅坐落在这片茂盛的西班牙橡树林的中央。住宅的中心是独具魅力的带有池塘的庭院，具有百年历史的房屋都矗立在四周。传统生活中的庭院，在现代户外生活中被赋予了新的意义。这栋住宅的设计充分体现了主人的殷勤好客。设计师在住宅中设置了宽敞的户外空间，还有一个带有顶棚的主通道，供休闲娱乐使用。起居室、餐厅、厨房三者形成了一个组合式空间，充分体现了这种休闲娱乐的主题。人们可以在这个开放的空间中享用美食，观看电视，闲聊小憩。厨房装备了各式顶级设施，为主人一家和客人提供餐饮美食。通过起居室巨大的滑动门可以直接步行至户外。娱乐之余可以在池塘边静静地放松，观看高尔夫球课程演示。西班牙橡树林中的住宅内部装修设计独特，古典式的石质装饰随处可见，设计师颇以此为傲。来自 Emerald Crest 的建筑商阿特·邓肯和斯科特·布莱恩特在这栋完美住宅的施工中，将他们的高超技艺发挥得淋漓尽致。

Lakeside Villa

湖滨别墅

DESIGNER | John Hathaway, AIA
DESIGN COMPANY | Vanguard Studio Inc.
LOCATION | 12910 Park Drive
AREA | 425 m²
PHOTOGRAPHER | Coles Hairston

The approach and entrance feature astonishing views of the lake and hills that surround the home. The tricking sound of a fountain invites you into the comforting entry court of this Italianate Lakeside retreat. The owners desired a private, cloistered retreat that could also used to entertain large groups. They turned to Vanguard Studio Inc. Vanguard Studio had a difficult time coming up with the complicated design plan. They had to design the house around the view, lake and sun. The client's privacy was an important factor, so coming up with a design that included a great view and privacy was quite complicated. The home features large windows to showcase the magnificent views of the lake. A groin-vaulted gallery welcomes visitors and contributes to create a special, intimately scaled and elegant passage that pulls you out of the lake. The bold arches echo throughout the house not only in the wall openings, but also in the ceilings. The clients required a state-of-the-art kitchen for entertaining that also has the same attention to architectural detailing as the rest of the home. Like in many of Vanguard Studio's homes, this house features a flowing living room, dining room, and kitchen; this creates a unique and open space. The main outdoor living areas are large enough to accommodate a crowd and sit literally on top of the lake. The main family room has a picturesque connection to the lake. The soaring family room focuses on the two-story fireplace spruced up with a relief of the "god of wind". The master bedroom has a breathtaking view of the lake. The master bathroom and its tiled soaking tub "with a view" are remarkable.

别墅位于湖畔，周边群山环绕，站在别墅的大门入口处，就可以欣赏到周边令人窒息的湖光山色。从不远处的庭院中传来喷泉悦耳的水声，仿佛在邀请来客进入这个典型的意大利式湖滨别墅。业主的最初构想是建造一个极具私密性的静谧住宅，同时也能承办一些大型的招待活动。于是，他们委托先锋工作室来进行设计。先锋工作室的设计师们经过长时间地精心规划，最终完成了这个十分复杂的设计方案。这个方案充分利用了住宅周边的群山、湖景以及自然光线。业主要求的私密空间是方案中最重要的元素之一，将这种高度的私密性与周边美丽的景色相互融合，是一件十分复杂的工程。设计师们为住宅设计了巨大的玻璃窗，可以观赏美丽的湖畔美景。交叉拱形的回廊体现着一种独特的私密感，走在优雅的通道里，随时感受外面美丽的风光。醒目的拱形设计随处可见，包括墙体上的拱门以及拱形的穹顶。应业主要求，厨房配备了顶级的设施，以备将来款待宾客。而业主对住宅其他各处建筑细节的要求也颇高。住宅内的起居室、餐厅以及厨房空气流通性很好，营造出一个独一无二的开放式空间，这是先锋工作室对于住宅设计的一贯风格。户外活动空间十分宽敞，足够容纳宾客团体在这里聚会活动，同时可以领略远处美丽的湖面景色。设计师们精心规划了主要的家庭活动室，透过巨大的窗户就可以直接欣赏外面美丽的湖光山色。家庭活动室中最醒目的就是两层楼高的巨大壁炉，威严耸立，仿佛是慰人心神的风之王者。主卧室同样可以直接看到美丽的湖景。主浴室里设计了平铺式的管式装饰，十分引人注目。

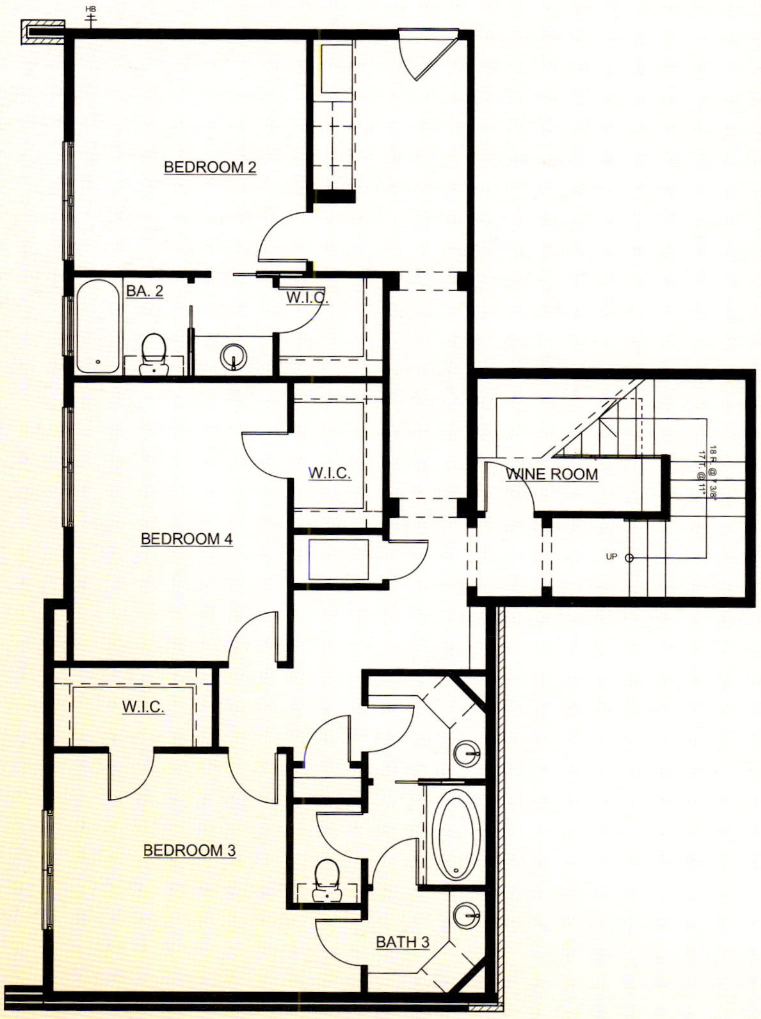

BASEMENT LEVEL

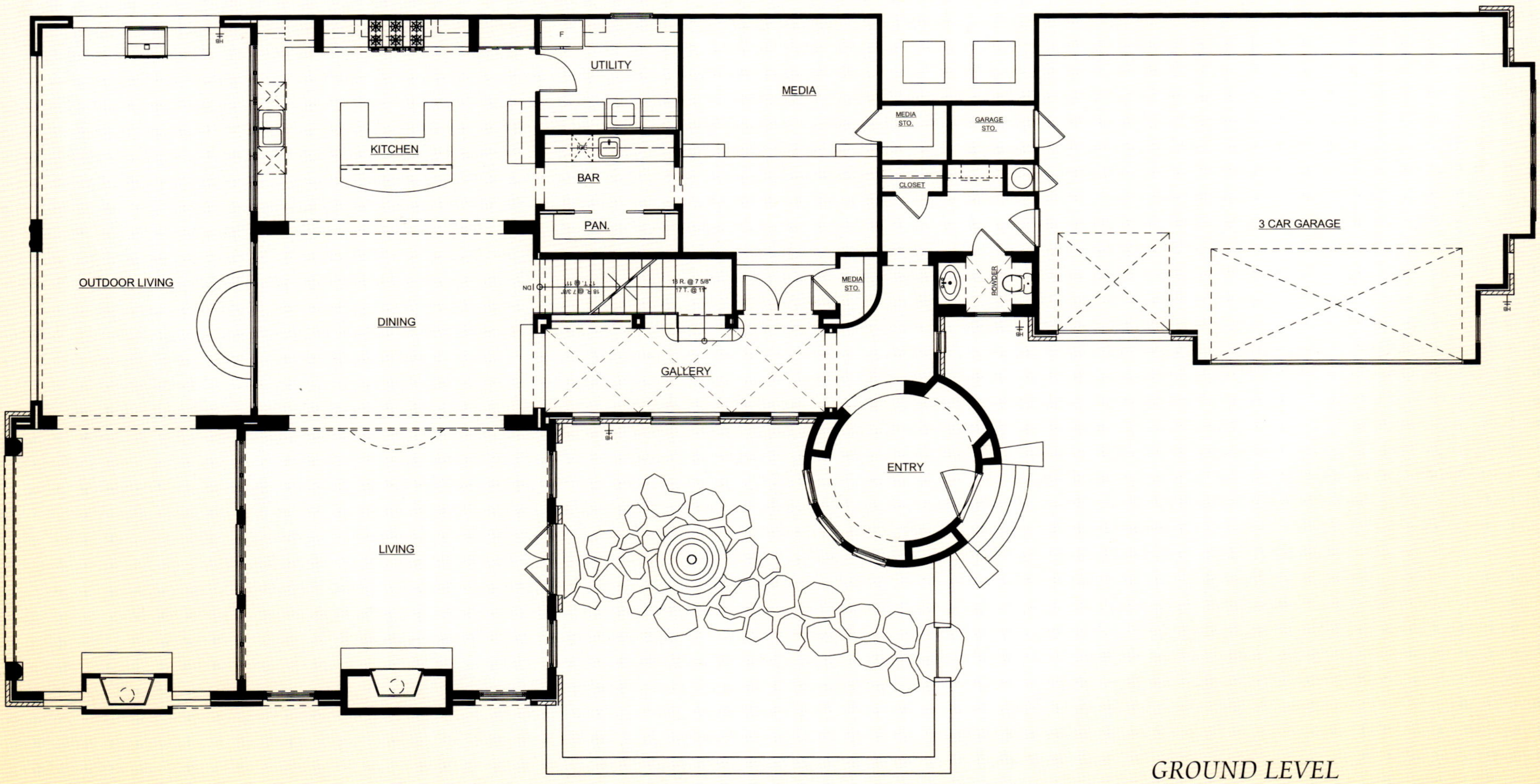

GROUND LEVEL

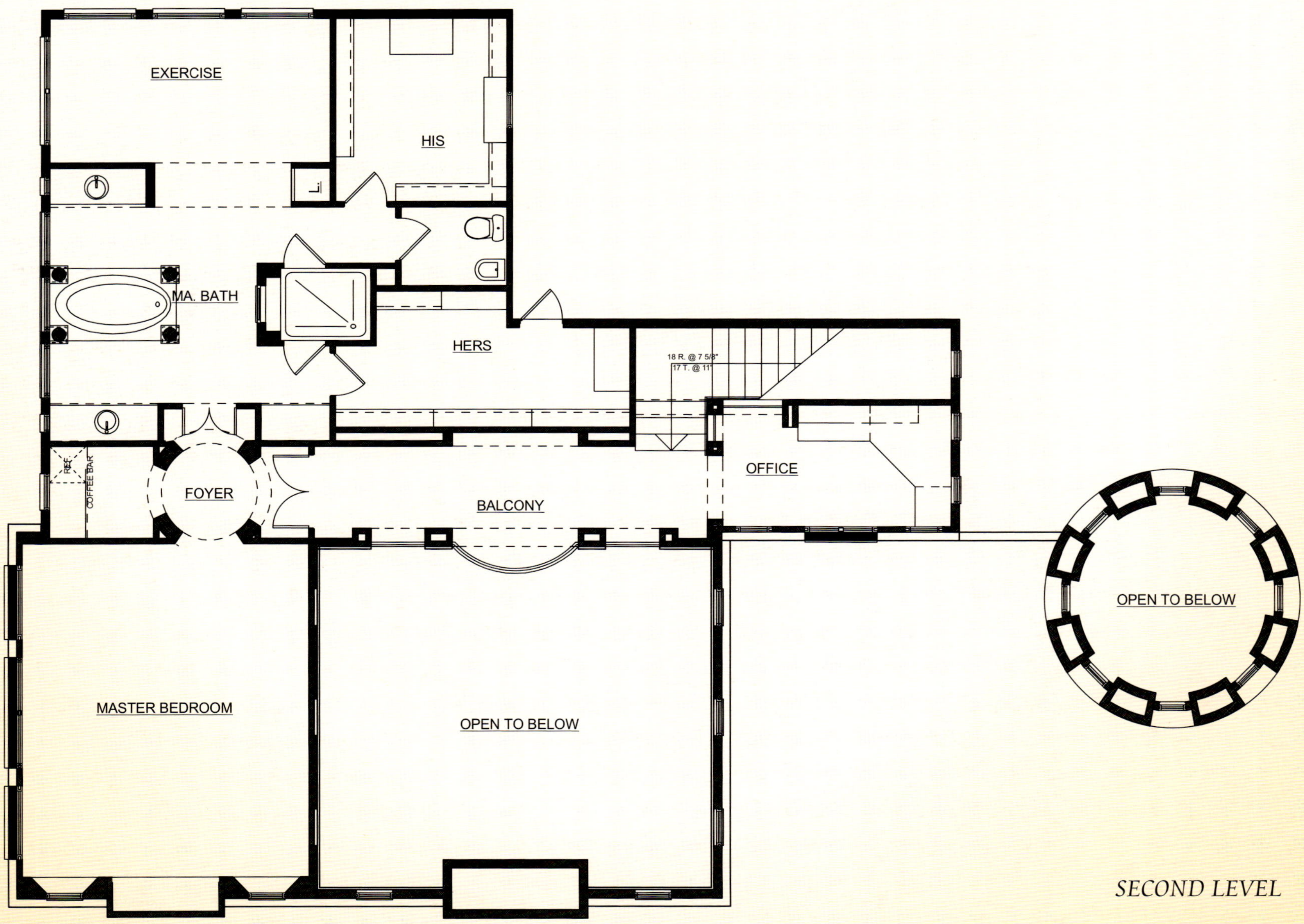

SECOND LEVEL

Arcadia Residence

阿卡迪亚别墅

ARCHITECT | Adele Chang
LANDSCAPE ARCHITECT | Robert Pressman, TGP, Inc.
FIRM | Lim Chang Rohling & Associates
LOCATION | Los Angeles, California, USA
AREA | 1,041 m²
PHOTOGRAPHER | Michael Arden

This Mediterranean villa, designed by architects of Lim Chang Rohling & Associates, is framed by old oak trees, sun-warmed in color and quietly elegant in its classic details and proportions. Though relatively large in square footage, the home still provides comfortably scaled spaces as a backdrop for daily living and family gatherings. It is organized into formal and informal wings which respect the formal axes of the home. The axial and symmetric design results in a sense of calm and order as well as easy flow of circulation. Indoor/outdoor connections are celebrated throughout with numerous arched French doors, which center on landscape vistas and create rhythm and interest inside and out.

The rear yard itself, with a long, narrow pool at its core, responds to the formal axes of the home that are established by the architecture. Rustic stone paving on the pool deck further contributes to the Mediterranean villa atmosphere. A raised spa with a broad spillway provides a focal point to be viewed from the family room. A series of low, colorful fountains with ornately decorated tile, on axis with the front door, create a rich and inviting entry experience.

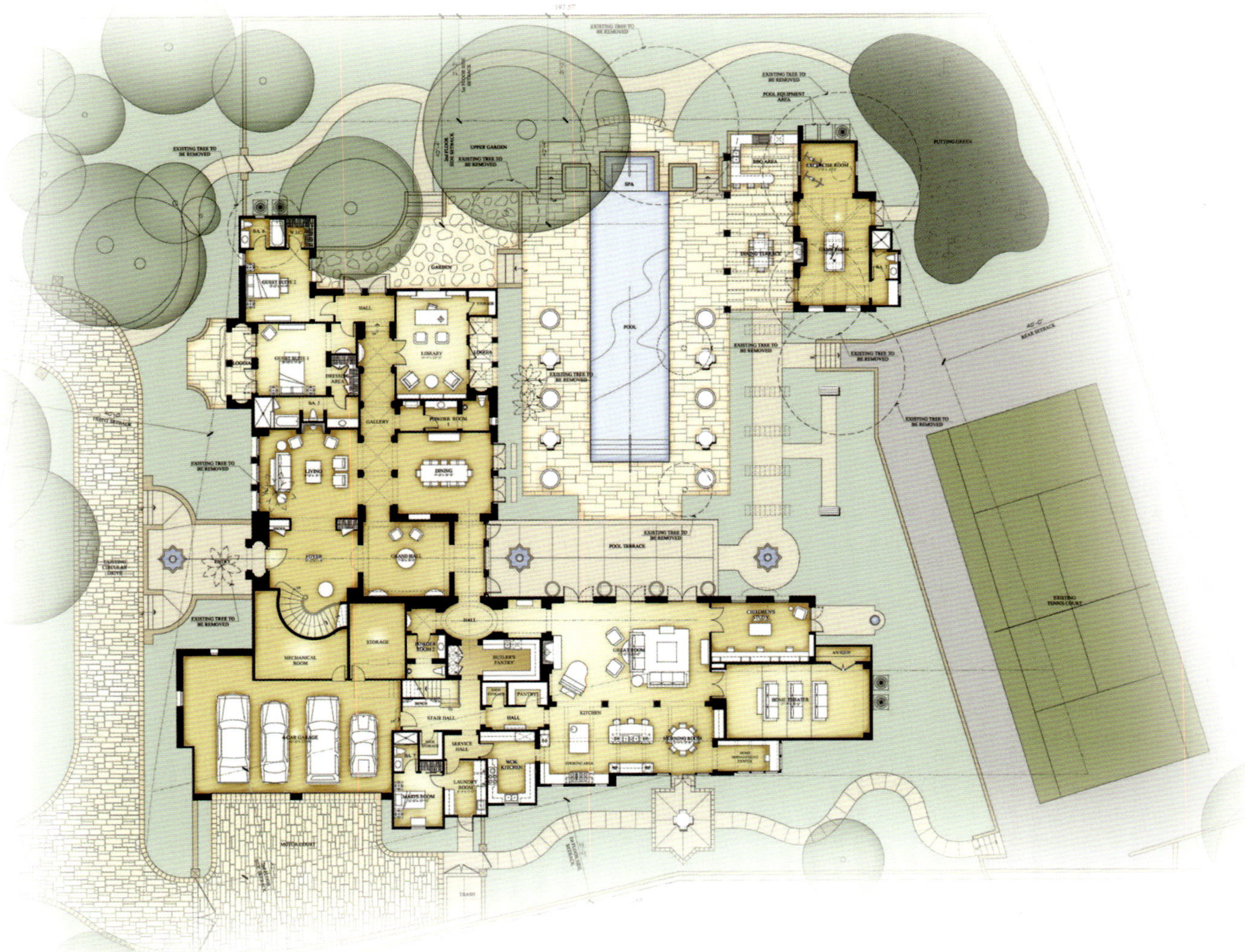

Glencoe Drive Elevation

阿卡迪亚别墅建筑具有典型的地中海风格，设计方案由林成·罗林设计联盟的设计师们共同设计完成。建筑的主体框架由陈旧的橡木构成，温暖的色调，简洁的线条，处处流露着古典建筑的传统细节和尺度比例。尽管建筑面积相对宽阔，还是为家庭的日常活动和聚会预留了足够宽敞而舒适的空间。这些空间在建筑中沿着轴线两侧分布，构成多种侧翼的空间形式。建筑整体沿着轴线对称设计，给人一种平静和规则的感受。多个法式拱门位于景观带的中央，作为室内外之间的过渡，舒缓空间节奏。

别墅还拥有一个独立的后院，中间是一个长条形的泳池，位于整栋建筑的轴线上。泳池边由质朴的石块镶嵌而成，具有典型的地中海别墅风情。从别墅房间可以直接看到后院中一个突起的水系景观，水流喷涌而下，是整个后院的亮点之一。前门附近的轴线上是一系列装饰华丽、五光十色的低矮喷泉，向宾客们展示着主人的富有和好客。

The Venezia

威尼斯住宅

ARCHITECT | Jon Rentfrow
FIRM | Rentfrow Design, LLC

This Mediterranean Style home stands grand and proud, both on the front and back. Sitting on a lake, most of the spaces orient towards the water on the back. The two-story, glass, stair tower is a very significant centerpiece to the overall design. Every space, in and out, of this home is ready for entertaining.

这栋住宅的正面和背面轮廓，都淋漓尽致地体现了典型的地中海风格建筑的宏伟与高贵。住宅坐落于湖畔，后面是波光粼粼的湖面。住宅整体分为两层，玻璃材质的楼梯间是整个设计装饰的亮点。建筑内外的每一寸空间无不带给人赏心悦目的享受。

LOWER FLOOR PLAN

3544 FinSqFt

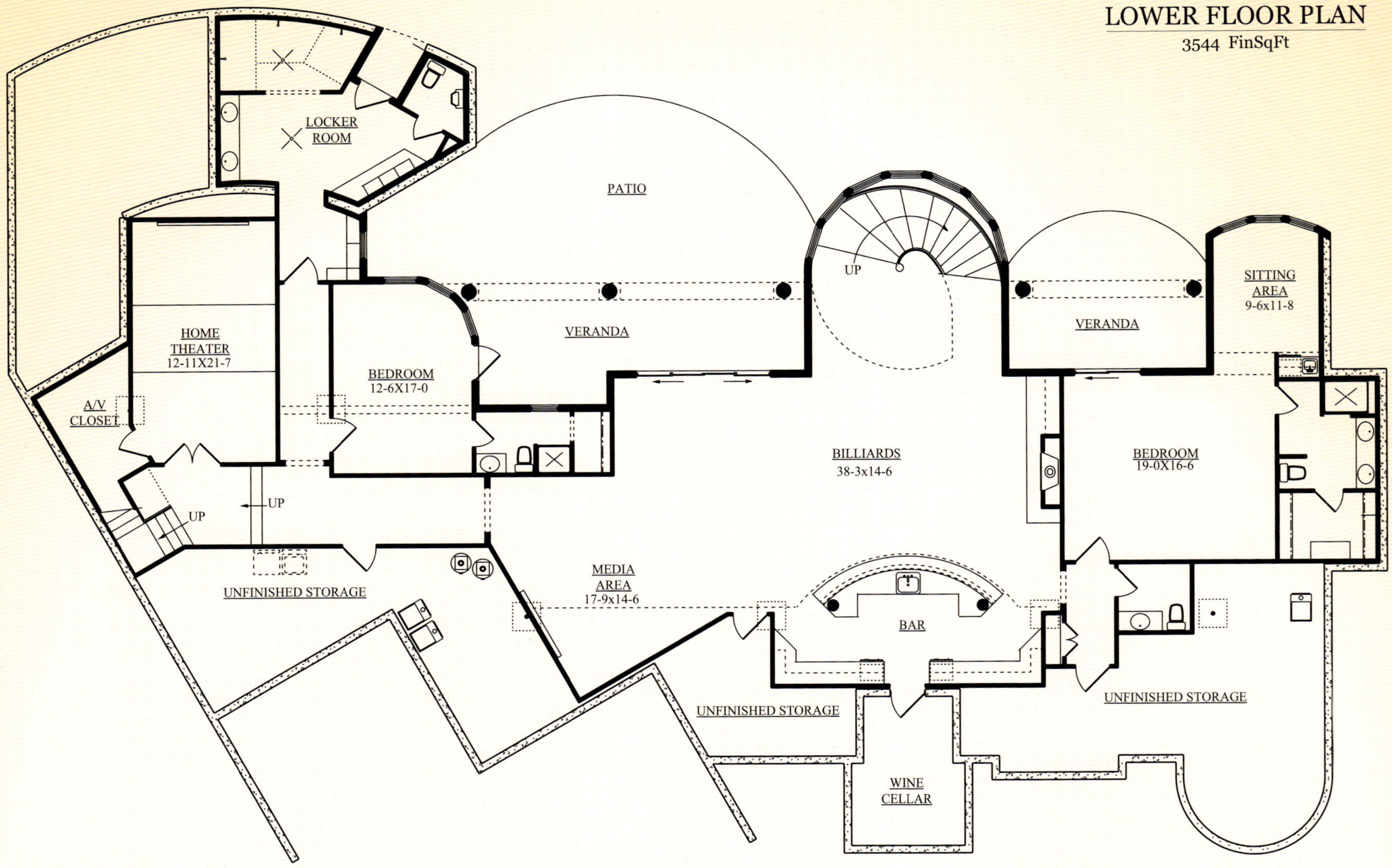

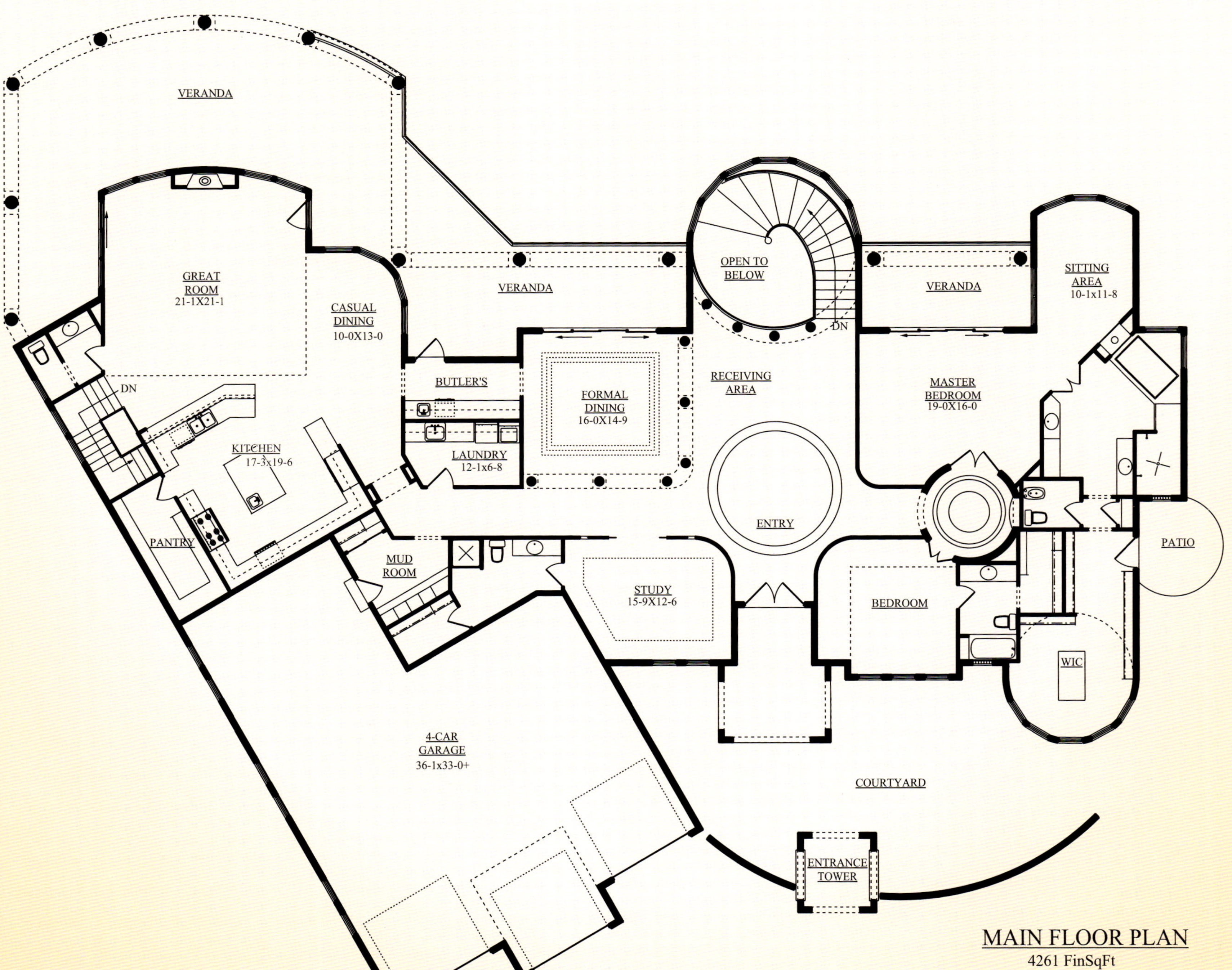

MAIN FLOOR PLAN

4261 FinSqFt

Kenwood Inn and Spa

肯伍德温泉旅馆

CONTRIBUTOR | *Kenwood Inn and Spa*

LOCATION | *Kenwood, California, USA*

One only has to peer through the front glass door of the Kenwood Inn and Spa to grasp its natural beauty. The 10,117 square meters grounds of the Inn are literally alive with oak groves, walnut trees and fruit trees including Meyer lemon, olive, pomegranate and fig. The Mediterranean-style villa has three lush courtyards – each with an exquisite saline swimming pool, fountains, a serene pond, meticulously landscaped flora, and abundant ivy cascading up, down, and around the Inn. Beyond Kenwood's vibrant, Euro-rich beauty, guests are treated to vineyard vistas which begin at the winery next door and end at the horizon.

In addition to the carefully selected wine list from the Inn's cozy bar, guests will find another level of wine enjoyment at Kenwood. The award-winning, Spa at Kenwood features the age-old European tradition of vinotherapy, offering rejuvenating and protective anti-aging skin treatments based on grapevine-derived extracts that have long been recognized by doctors and followers alike. The Spa specializes in body treatments including a wine bath, wrap, scrub and massages.

The restaurant at the Kenwood Inn and Spa offers yet another compelling draw. The dining room, kept warm with a fireplace on cool nights, and open to the courtyard when weather permits, is for guests only. Each morning, guests find a complimentary breakfast offering hot dishes to order, local fruit, house baked pastries and specialty coffee. Executive Chef Steven Snook utilizes seasonal fruit produced from the property's fruit trees and garden, and shops the Sonoma Valley's best farms and farmer's markets to stock the kitchen with in-season products to create his artisan Italian menus. Guests may also consider lunch or dinner on the patio, or in the intimate wine bar as dining options.

坐在肯伍德温泉旅馆中，透过前方巨大的玻璃门就可以欣赏这里美丽的自然风光。温泉旅馆占地约 10 117 平方米，坐落在一片树林中，这里有橡树、胡桃树以及野柠檬、橄榄、石榴、无花果等果树。整个旅馆别墅为地中海风格，设有三个不同的院落，经过精心设计，每座庭院都有一个盐泉泳池、喷泉和池塘，周边是郁郁葱葱的植物群落，大量的蔓藤植物从墙上垂悬而下，围绕在别墅周围。在这里，游客们除了可以欣赏充满欧陆风情的美景，还可以沿着葡萄藤架下的小路，从邻近的葡萄酒酿造作坊，一直徜徉到远方。

游客们可以在旅馆温馨的酒吧中品尝经过精心挑选的各类葡萄酒，还可以在这里享受一种不同寻常的温泉体验。肯伍德温泉旅馆别出心裁地推出一项在欧洲由来已久的传统葡萄理疗体验，这是一种多年来经过医学专家们普遍公认的理疗方式，使用葡萄藤中提取的营养成分对肌肤进行护理，达到抗衰老、焕发青春的功效。整套理疗体验包括洒浴、缠绕、擦洗和按摩等程序。

温泉旅馆的餐厅可以为游客们提供各种难以抗拒的美食享受。入夜，餐厅利用传统的壁炉取暖，为游客提供一个温暖舒适的就餐环境。如果天气条件允许，还提供室外就餐空间。早上游客们可以免费享用精美的早点，包括热菜、当地水果、新鲜烤制的糕点以及味道独特的咖啡。行政总厨史蒂文·斯诺克先生从自家花园和附近索诺玛山谷最好的农场中选用当季的新鲜水果和蔬菜，为游客们提供各式意大利风味菜式。当然，游客们也可以选择在院子或葡萄酒吧中享用午餐和晚餐。

KENWOOD
INN
&
SPA
RECEPTION

THE SPA
AT
KENWOOD INN
est.
1994

CONDE NAST JOHANSENS
RECOMMENDED

Cimarron Hacienda

锡马龙庄园

DESIGNER | John Hathaway, AIA
DESIGN COMPANY | Vanguard Studio Inc.
LOCATION | 109 Spearpoint Cove, USA
AREA | 520 m²
PHOTOGRAPHER | Coles Hairston

This stunning example of Spanish architecture is full of unique iron, stone, and wood and arched details. The Spanish style home provides lots of outdoor space for entertainment and relaxation. The outdoor space also extends to the second floor of the home, where the family or guests can take in the amazing views of the country side. The second floor also overlooks the pool so people can watch the entertainment below. The upper veranda is by far the most unique part of this home. It is an unusually large outdoor space for relaxing and entertaining. The driveway courtyard provides a place where the owners can park their cars without anyone seeing them. Another unique feature of this Spanish home is the detached guest room. The guest room gives guests their own private space where they can live.

109

锡马龙庄园是一处令人叹为观止的建筑。独特的铁艺、石雕、木质结构和拱形结构，每一处细节无不诠释着典型的西班牙式建筑风格。西班牙式建筑有很多供人们娱乐休闲的户外空间的设计，这些户外场所的构造可以延伸到建筑的二层，家人或者是客人可以在那里一览当地的乡村风景。在建筑的二层可以俯瞰整个游泳池，这样即便在二楼也可以和泳池边的人们一起共享欢乐。上层的阳台更是这座建筑最独特的地方，为人们的休闲娱乐提供了更大的户外空间。主人可以把车停在庭院式的车道上，极具私密性。这座西班牙式建筑的另一独特设计就是主客分离，客房很好地保证了客人在此居住的私人空间。

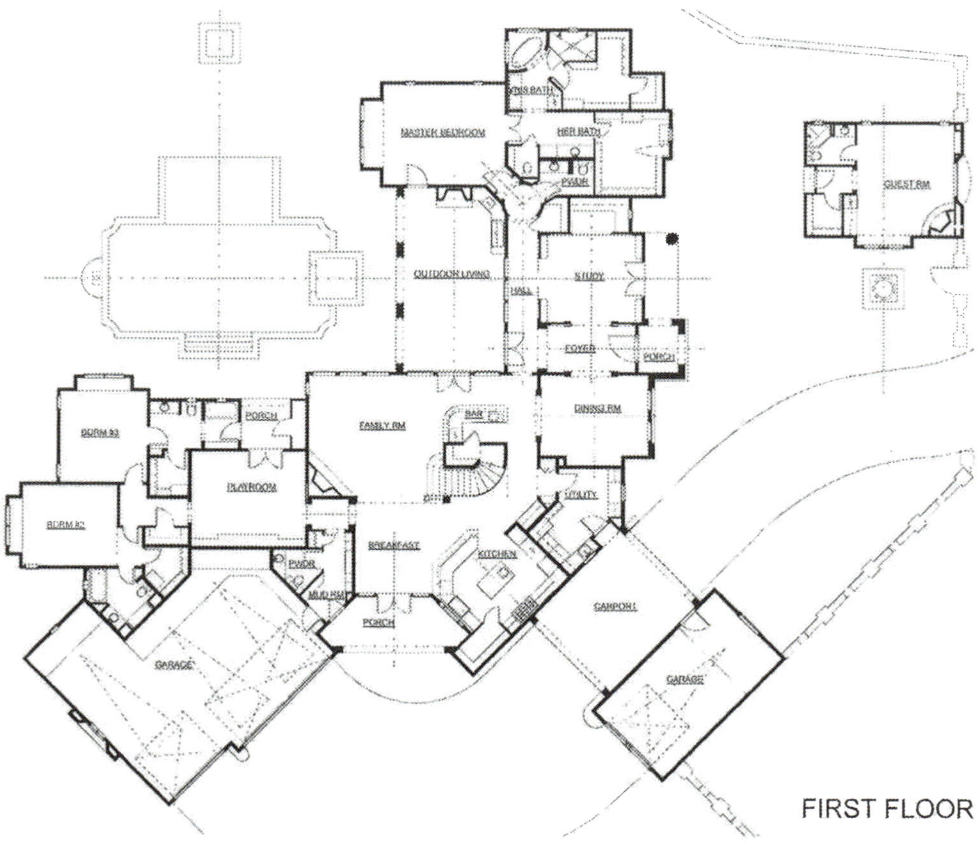

FIRST FLOOR

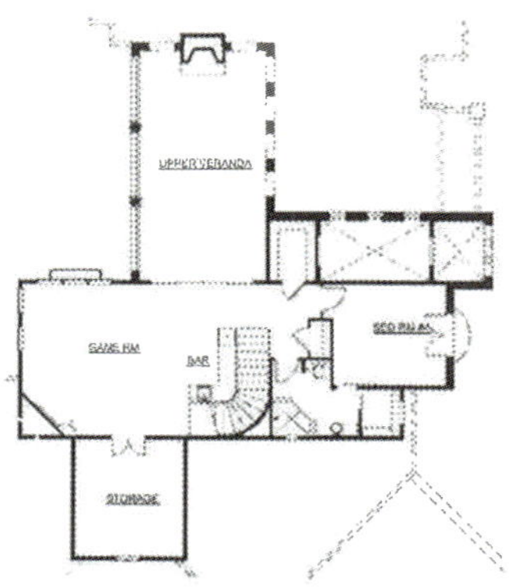

SECOND FLOOR

DANOS EL PAN NUESTRO DE CADA DIA

Southpointe

索斯庞特住宅

INTERIOR DESIGNER | Deborah Costa
DESIGN COMPANY | StageCrafters
ARCHITECT | Gary Randall Shipman
LOCATION | Sacramento, CA, USA
AREA | 409 m^2
PHOTOGRAPHER | Trevor Jobson

A distinct residence is with over 409 square meters of living space. The home features 4 bedrooms, 4.5 bathrooms, formal Living and Dining Rooms, Family Room, Office, Game Room and 6 car garages. The open floor plan is ideal for entertaining with a large gourmet kitchen with island and bar counter seating and oversize living areas overlooking the backyard.

LIFE IS A GIFT

这是一个独特的居所，拥有超过409平方米的居住面积。房间独特地设计了4个卧室、4个浴室、一个半卫（只有洗手盆和马桶）、正式的起居用餐房间、家庭活动室、办公室、游戏室和6个车库。为款待客人品尝美食，采用了开放式的地面设计，并设计了超大的厨房空间，里面设有岛台和吧台座椅，另外还有超大的活动区域，在这里还可以眺望整个场院。

Ercolano Villa

埃尔科拉诺别墅

DESIGNER | Fabrizia Frezza
DESIGN COMPANY | Fabrizia Frezza Architecture & Interiors
LOCATION | Herculaneum, Naples, Italy
AREA | 465 m²
PHOTOGRAPHER | Guido Fuà

The villa which dates back to the 1970s had an excellent structural base on which to build. The plan of the ground level did not need substantial changes, however, it needed to be updated. A dark and disharmonious atmosphere existed due to the use of tiles, stonework, wood frames, custom-made wooden furniture, fabrics and decorations. The interior of the villa did not reflect the exterior. The white stucco and the form of the villa denoted a typical Mediterranean style. The main objective of the architect's project was to create harmony between the interior and exterior. The architect brightened the ambiance with the use of white waxed terracotta on the ground floor. The upstairs bedroom has an overall warm design. The bleached oak wood floors blend beautifully with the use of candid white warm wood. The window frames have been bleached and modified to let as much sunlight as possible filter through.

The harmonious Mediterranean tones, the pastel-colored linens, the deep purple velvet sofas are the perfect backdrop for the family's antique furniture and the architect's exclusively designed pieces. Selected family furniture, as the armoires and a large dresser were rediscovered, whitened and finely decorated with gold patina creating a harmonious retro feeling.

The elegance and simplicity of candour white predominates in all its forms including in the baths. The stone sinks are set off by modern stainless steel accessories.

The dominant white terrace contrasts with the pure blue sky and sea which reflect into the swimming pool. Dark wooden patio furniture together with white built-in chaises lounges frame the pool. The view of Mount Vesuvius which dominates the city of Herculaneum becomes surreal.

SECTION E-E'

SECTION F-F'

SECTION G-G'

埃尔科拉诺别墅始建于20世纪70年代，堪称当地建筑史上的一个杰作。在改建规划中，别墅的地面基本不做太大的改动，只是做一些适时的更新改造。原有的建筑中采用了一些暗色元素，如瓦片、石雕、木架、定制的实木家具、编织物以及装饰图案，形成了一种与整体风格不搭配的氛围。别墅的内部风格与外部完全不同。白色的粉饰泥灰墙体是典型的地中海风格的象征。设计师们的主要目的是在别墅内部和外部空间之间达到一种和谐平衡的效果。采用白色陶土瓷砖铺设地面，使得周围环境变得更加明亮。楼上的卧室设计则侧重于温馨舒适的风格。地板是由白色的暖木和浅色的橡木地板铺成，窗框使用淡色实木，窗体宽大，保证了室内充足的阳光。

和谐动人的地中海建筑基调，色彩鲜明的线条，深紫色的天鹅绒沙发，搭配仿古风格的家具以及设计师精心设计的装饰，共同形成了一个轻松休闲的生活环境。那些精致的家具，如宽大的衣橱和梳妆台，白色基底上搭配着带有铜绿的金色配件，给人一种时光倒流的感觉。

白色的主色调显得优雅而又质朴，甚至连浴室也延续了这种特色，浴缸由精致的石块砌成，配有现代风格的不锈钢配件。

白色的露台上面是蔚蓝的天空，远方是深邃的大海，这一切都倒映在庭院里的池塘中，色彩鲜明，十分动人。池塘边上的庭院里有深色实木家具和白色的休息大厅，远处是维苏威火山下的赫库兰尼姆市，在夕阳的映衬下，构成了一幅超现实主义的风景画。

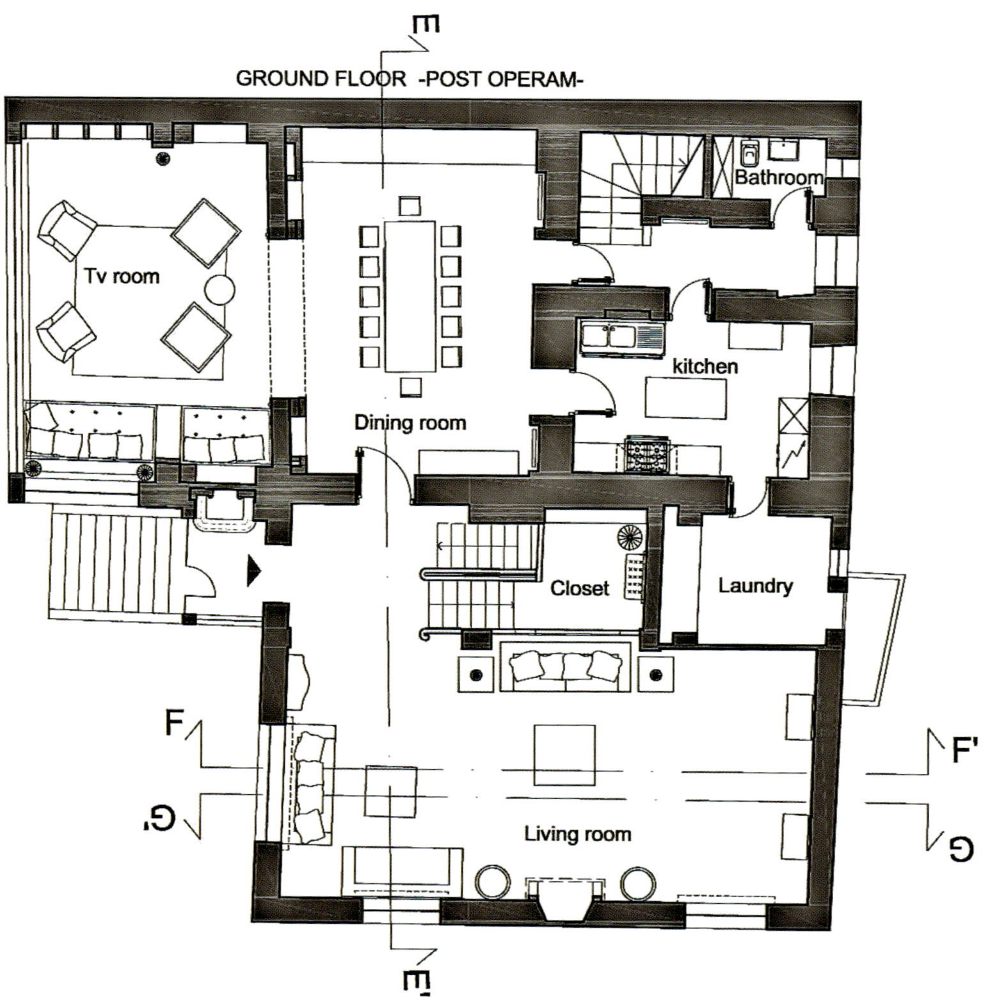

GROUND FLOOR -POST OPERAM-
Tv room
Dining room
Bathroom
kitchen
Closet
Laundry
Living room
E
E'
F
F'
G
G'

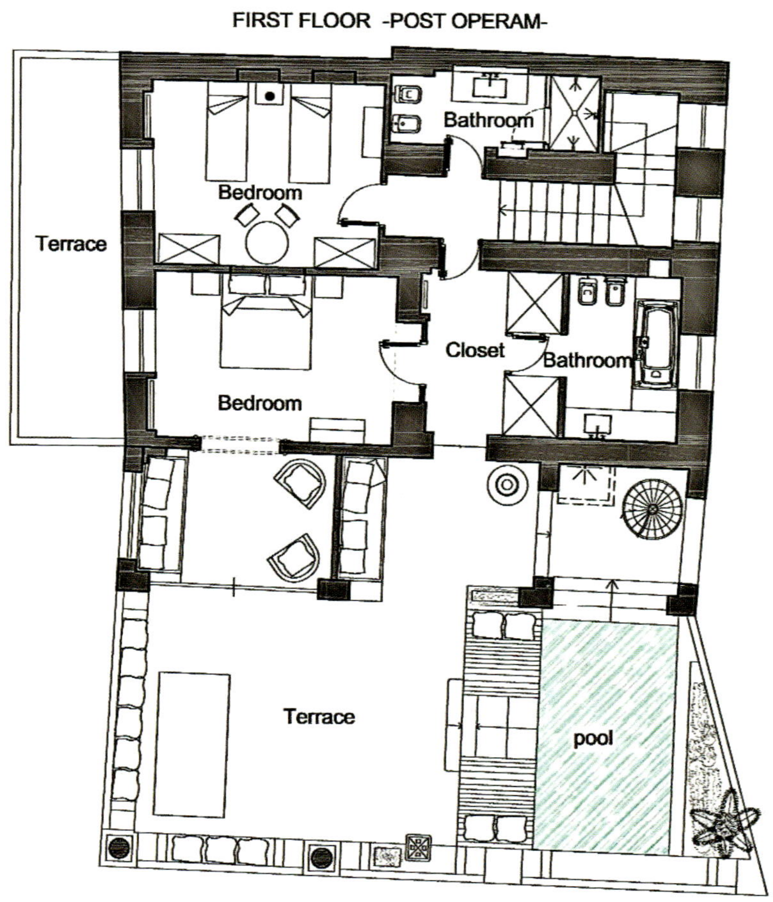

FIRST FLOOR -POST OPERAM-
Bedroom
Bathroom
Terrace
Closet
Bathroom
Bedroom
Terrace
pool

Casa H

H住宅

ARCHITECT | *Federico Martinez Garcia*
FIRM | *Constructora Malver*
PRESIDENT AND OWNER | *Miguel Angel Arce*
DIRECTOR OF CONSTRUCTION | *Carlos Lira Alvarez*
DIRECTOR OF DEVELOPMENT | *Bernardo Arce Valencia*
LOCATION | *Los Cabos, Mexico*
PHOTOGRAPHER | *Rigoberto Moreno*

The combined work that the builder and the interior designer must do is of the essence. The satisfaction of the clients' needs will depend on their joint understanding and teamwork. In , the symbiosis between MALVER and the interior designer was total. It bears to mention that the interior designer herself was the owner; clear, professional, and with creative proposals in all instances. If the Company finds pleasure in serving its clients, the collaboration with Mrs. Heltne was a delight, which resulted in a magnificent residence.

"In working with Malver through the design, bid and construction phases we were extremely pleased with their professionalism, timeliness and finished result."

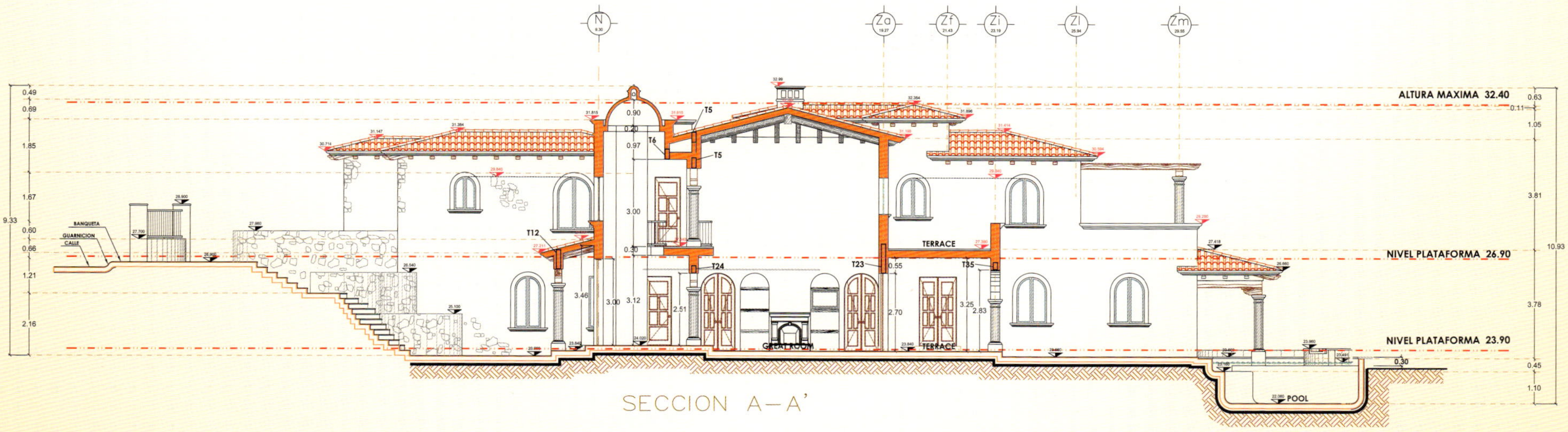
ALTURA MAXIMA 32.40
NIVEL PLATAFORMA 26.90
NIVEL PLATAFORMA 23.90
TERRACE
POOL

SECCION A—A'

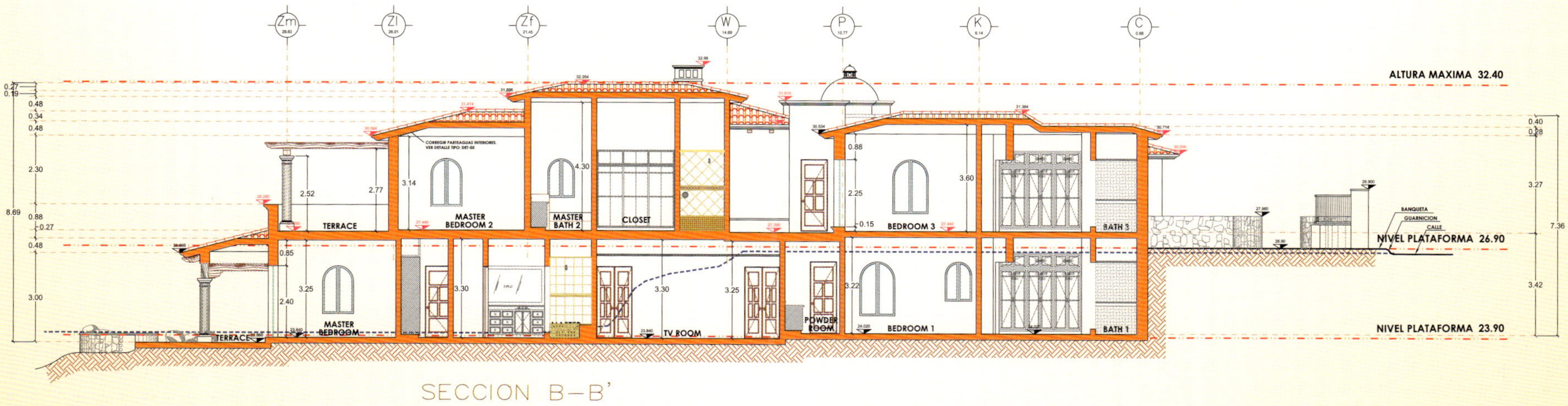
ALTURA MAXIMA 32.40
NIVEL PLATAFORMA 26.90
NIVEL PLATAFORMA 23.90
TERRACE
MASTER BEDROOM 2
MASTER BATH 2
CLOSET
BEDROOM 3
BATH 3
MASTER BEDROOM
TERRACE
TV ROOM
POWDER ROOM
BEDROOM 1
BATH 1

SECCION B–B'

建筑设计师和室内设计师的合作成果体现了整栋建筑的精华所在，同时，他们对于设计方案的理解以及团队合作，是满足客户需求的基础。整栋建筑充分体现了建筑规划设计与室内设计的契合与共生。值得一提的是，室内设计师就是住宅的业主，她在所有的室内设计中都清晰地表现出一种简洁、专业并富有创新性的概念。如果建筑设计公司以服务于客户为乐事，那么与海尔特尼太太的合作无疑是一个令人愉悦的过程。他们通力合作，共同建造了一栋高贵华丽的居所。

海尔特尼太太认为，在与马维尔公司合作的整个过程中，包括设计、投标以及建筑等各个阶段，对于他们的职业水平，按时完成工期的时效性以及项目完成的质量都感到十分满意。

McMonigle

麦克莫尼格

ARCHITECT | *Robert Hidey*
FIRM | *Robert Hidey Architects*
LOCATION | *Irvine, CA, USA*

Situated on a prominent site at the entry to the affluent neighborhood of Shady Canyon, in Orange County, California, the Ranch House evokes the Southern California of the 1920s, when homeowners were invested in the fantasy of being gentleman farmers, living on a hacienda. With the Ranch House, Robert Hidey Architects turns the vernacular of the Mexican "rancheros" into something elegant and timeless.

The 557-square-meter home is an unabashed homage to Wallace Neff, one of the acknowledged masters of the American "Mission Style". Inspired by the spirit of his homes, the firm exhaustively studied Neff's work, utilizing the classical principles of axis, view corridor and focal point in this home.

The timeless grace and beauty evident in the Ranch House were born from a clever solution to a challenging site. In addition to a mandatory one-story height requirement by the City, the setback lines created a unique triangular pad, surrounded by two streets, with a 4.27-meter incline.

The solution was a house and roofline that gracefully stepped down to follow the topography. A grid system was used to lay out the disciplined floor plan. Exterior walls create courtyards that enjoy sun and shade at different times of the day. Adobe brick simulates an aged, weathered appearance. A roofed loggia creates a low silhouette with surprisingly high ceilings inside; massive, authentic, old-world ceiling beams poke through exterior walls at 1.22-meter intervals to support the eaves. The home's tower features a private exterior stair leading to an entry into the office/studio above the foyer, while below, a connoisseur's wine cellar is used for entertaining guests. An authentic 300-year-old Mission Bell functions as a doorbell.

With its symmetry, quiet palette and soft natural light, the Ranch House has the ambience of serenity. Hand-worked stucco walls, deep loggias, hammered ironwork and adobe brick are authentic details that recall the glorious "rancho" days when artistry and craftsmanship were everything.

43

Motorcourt
Garage
Dn
Up
Garage
Laundry
Her Clo
Her Clo
Her Bath
Dn
His Bath
His Clo
Loggia
Family Rm
Kitchen
Nook
Sitting
Dn
Butler's Pantry
Loggia
Master Bedroom
Loggia
Courtyard
Dining
Loggia
Pwd
Foyer
Living
Bedroom 2
Up
Ba 3
Bedroom 3
Courtyard
BLUE HERON

本案位于加利福尼亚州的奥兰治县，地理位置十分优越，毗邻著名的富人聚居区绿荫峡谷的入口。本案建筑相对比较低矮，充分体现了一种20世纪20年代南加利福尼亚地区的特色，那个时候房屋业主们通常希望成为有身份的农场主，因而投资建造庄园住宅。罗伯特·西迪设计室的设计师们在这些低矮的建筑中加入了墨西哥农场的建筑元素，将之改造为集优雅与坚固为一体的乡村住宅。

建筑占地557平方米，参照了公认的美洲布道院风格建筑大师华莱士·奈夫的作品风格。为了从华莱士·奈夫的作品中获得灵感，设计师们研究了华莱士·奈夫的所有作品，在建筑设计中采用了古典建筑中常用的建筑轴线、观光走廊以及建筑中心等概念。

为了遵循城市规划有关单层建筑的规定，同时将优雅与美丽的元素引入农场风格浓郁的建筑中，设计师们巧妙地利用建筑选址解决了这两个难题，利用两条街道和一条4.27米长的斜坡，构成了一个独特的三角形建筑地块。

房屋建筑和屋顶的轮廓自然优美地沿着地形向下延伸，在建筑的平面图上形成了一种网格式体系。利用外墙围出一个庭院，白天可以在这里尽情享受和煦的阳光。风化斑驳的墙砖仿佛经历了时光的洗礼。带有天棚的凉廊画出一道低矮的轮廓，每隔1.22米就有一道贯穿外墙的木梁支撑屋檐，质朴、厚重，带有几分古意。另外还有一个楼塔，外部设有楼体，直通上面的办公室，下面是为客人准备的极具品位的藏酒室。一个具有300年历史的教堂大钟充当了门铃的角色。

建筑外形十分对称，加上静谧的色彩和柔和的自然光线，共同构造了一个宁静祥和的居住环境。人工涂抹的灰泥墙体、长长的凉廊、百炼成钢的铁艺以及土坯砖式构造等每一个细节都体现着艺术与技艺的水平，充分展示了大牧场时代的辉煌。

Vizcaya at Jinqiao

金桥维兹卡亚会馆

ARCHITECT | *DDG, Baltimore, Maryland, USA*

LOCATION | *Shanghai, China*

AREA | *111,722 m²*

Vizcaya at Jinqiao brings the relaxed atmosphere and lush feel of a Mediterranean hill town to Eastern China. This private community outside Shanghai targets an upscale clientele looking to experience the luxury and convenience of a self-contained residential development replete with modern amenities and convenient to local schools and commercial districts.

Creative land planning transforms a relatively flat site into a tableau of gently rolling hills, palms, and villas. Private residential courts, individually designed and themed, feed into a central avenue linking to the project's centerpiece, a spectacular community clubhouse. The clubhouse, surrounded by community gardens and sports facilities, contains a full service health spa and restaurant facilities, as well as a combination of indoor/outdoor heated pool. Clay tile rooftops, white stucco exteriors, towers, cupolas and indoor and outdoor exposed wooden beams establish a strong thematic identity, with palm-tree-lined streets and courtyards contributing to a tranquil, serene neighborhood sensibility.

The community, utterly unique to this part of the world, offers a wide selection of occupancy models and home designs, with 190 units ranging from three to five bedrooms. The convenience to Shanghai, security, and intimate feelings of community make Vizcaya a highly desirable residential destination.

D1
SD1
SD4
SD4
SD1
SD2
SD2

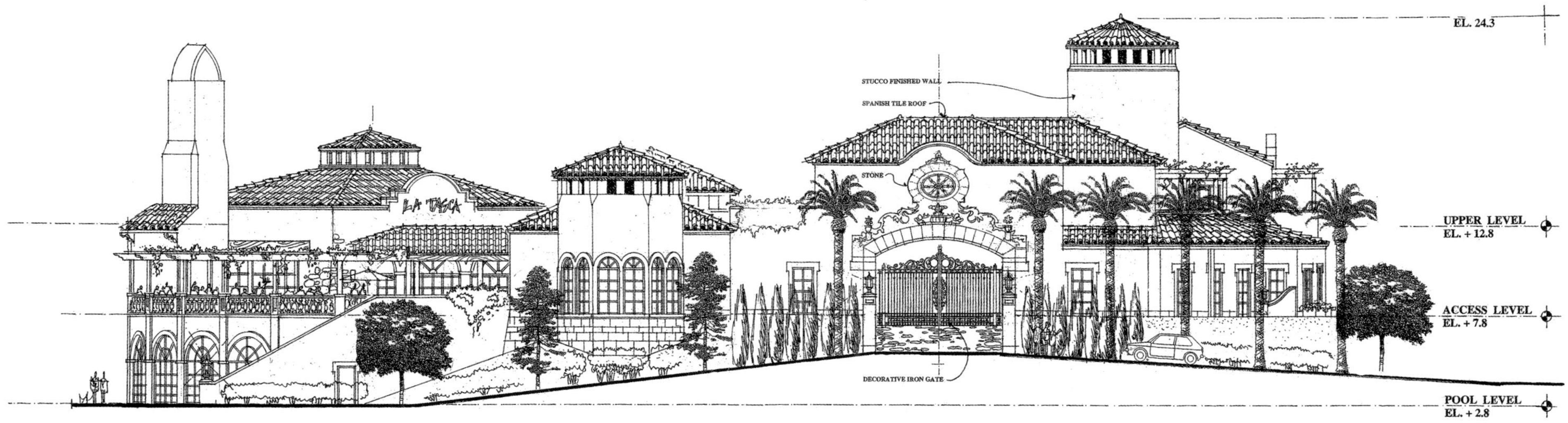

2 SIDE ELEVATION (EAST)
SCALE= 1:100

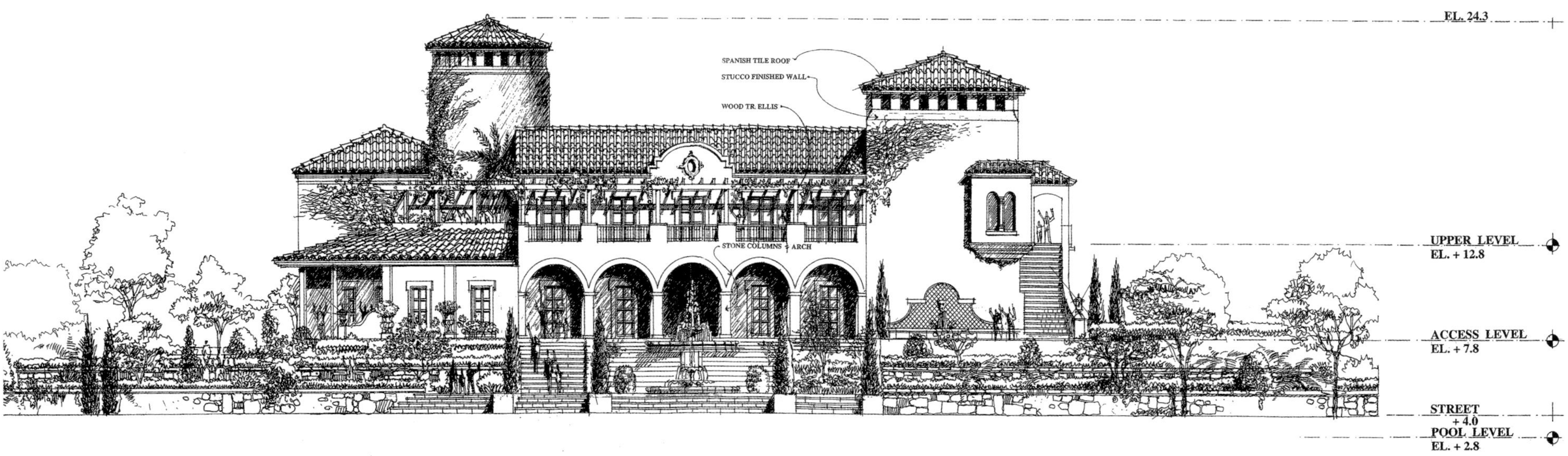

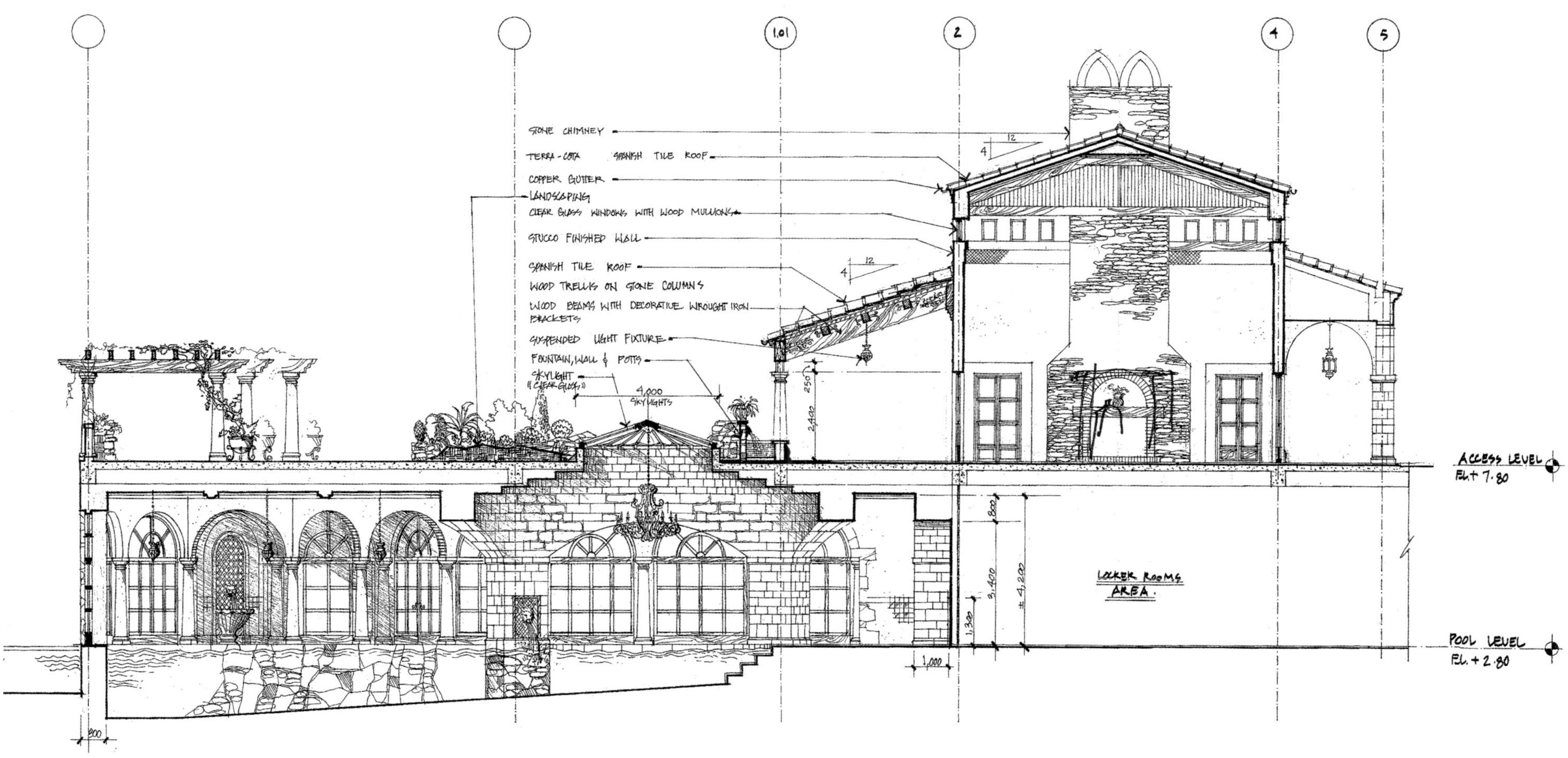

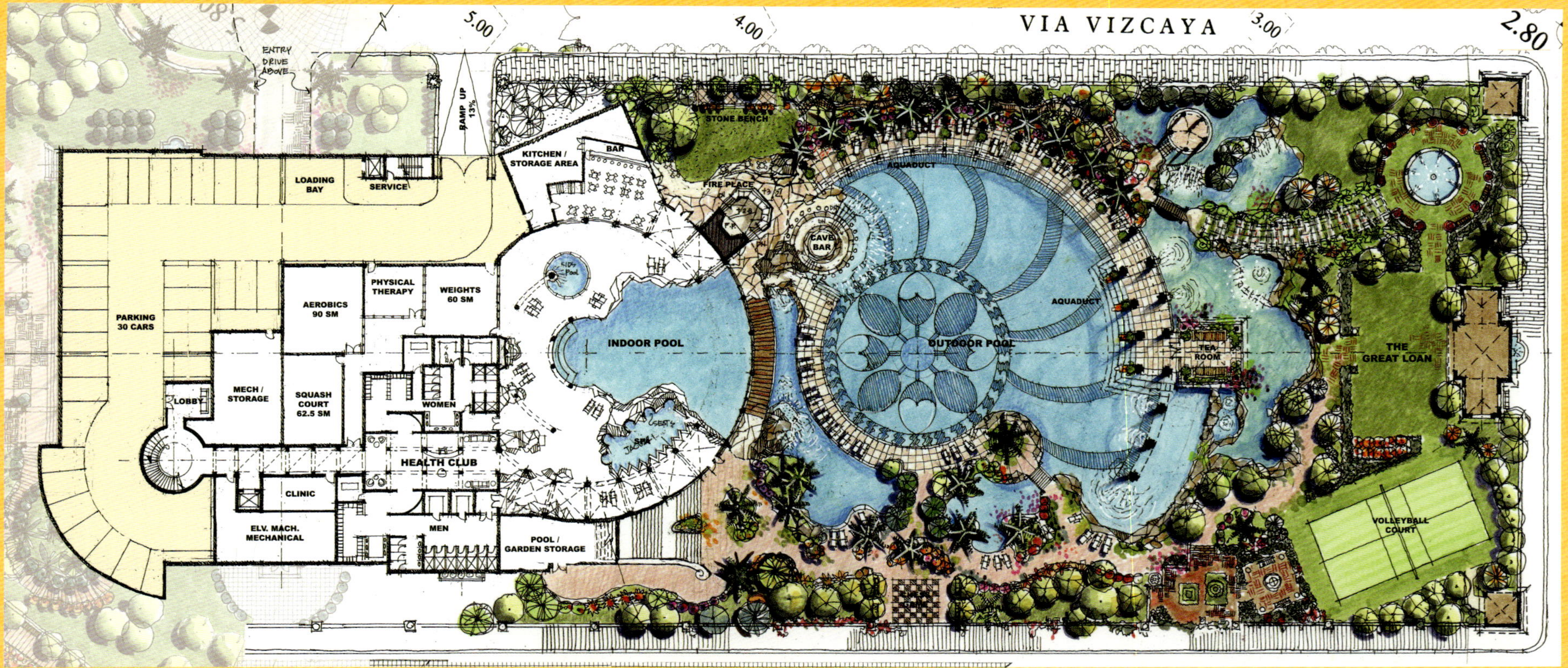

CLUB HOUSE POOL LEVEL PLAN
SCALE 1 : 100

CLUB HOUSE ACCESS LEVEL PLAN
SCALE 1 : 100

CLUB HOUSE UPPER LEVEL PLAN
SCALE 1 : 100

维兹卡亚会馆位于休闲幽静的金桥别墅区，这里是位于中国东部的一个具有典型地中海风格的山坡小镇。这个私密的社区专门为来自上海市区的高端客户打造，不仅提供奢华高贵的居住环境、便利齐全的生活设施，周边还配备了现代化的学校和商业区。

别具心裁的创新性规划，将这一片相对平坦的地块打造为一个极具个性的社区，这里的小山、棕榈树和别墅都充满了西班牙建筑的特色。每一栋私人别墅，都经过单独的设计，具有特定的风格，最中间是林荫道，直接通向华丽壮观的社区俱乐部。俱乐部周围分布着社区内部的花园和体育设施，包括一个健康理疗服务中心和餐饮中心以及一个室内外恒温游泳池。黏土瓦片屋顶、白色灰泥墙体、塔楼、瞭望塔和室内外裸露的实木房梁，共同创造了一种强烈的舞台剧场效果。街道和庭院两边是高高的棕榈树，营造出一个静谧安宁的居住环境。

金桥社区与周边社区千篇一律的风格完全不同，可以为顾客提供各式各样的装修模式和风格，一共190个住宅单元，每个带有三至五个卧室。社区毗邻上海市区，安全性高，私密性强，是一个十分高档的宜居社区。

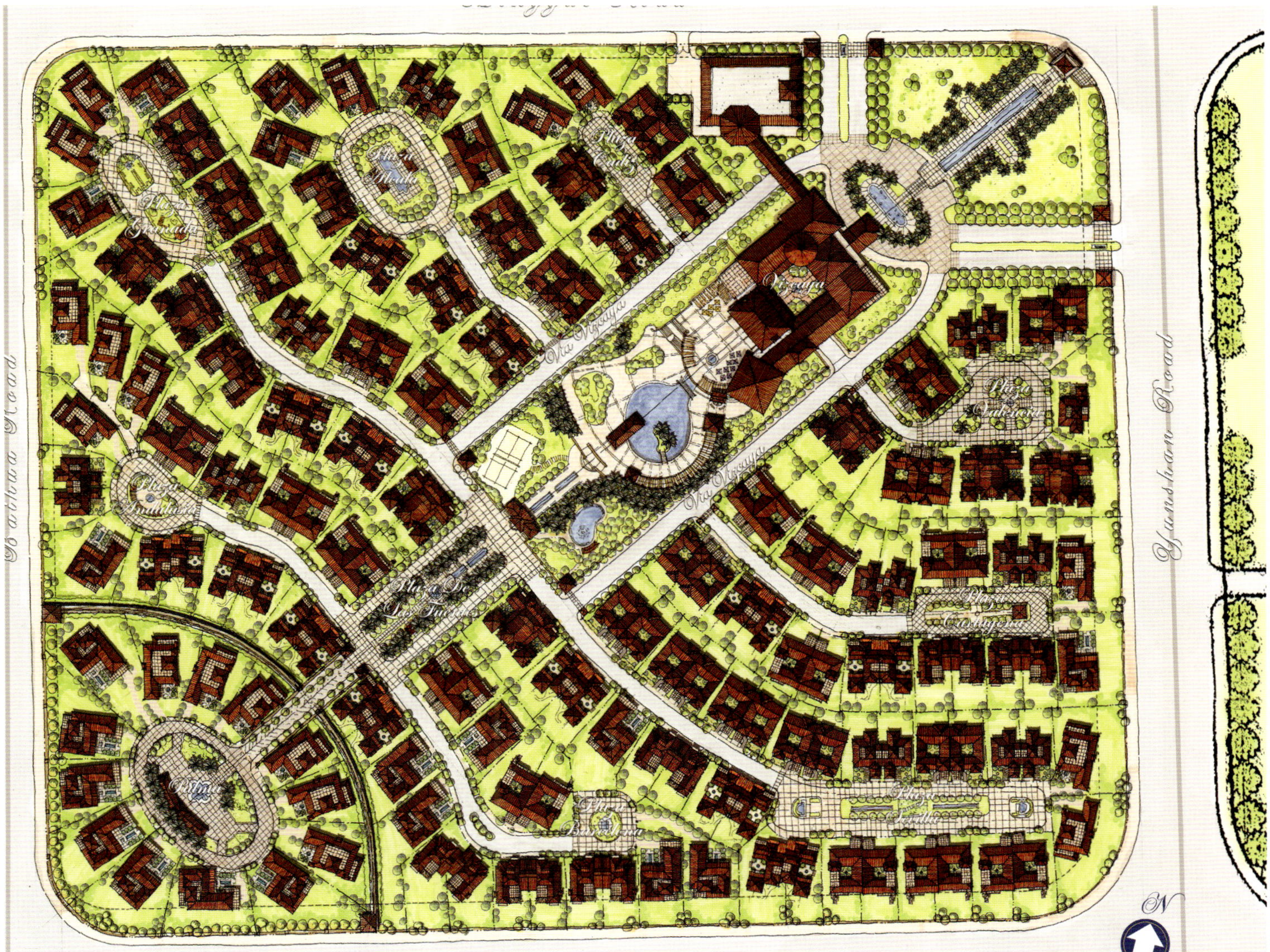

Anacapri Villa

阿纳卡普里别墅

DESIGNER | *Fabrizia Frezza*

DESIGN COMPANY | *Fabrizia Frezza Architecture & Interiors*

LOCATION | *Anacapri, Naples, Italy*

AREA | *380 m²*

PHOTOGRAPHER | *Guido Fuà*

The villa is exposed to both the north and the south. The kitchen, dining room and living room form a "unicum" with the exterior space through the use of large glass windows and the same flooring throughout, making the area become one and doubling the 75 square meters. The living area looks onto the terrace which becomes an integrated part of the home. Built-in, sinuous seating, covered with cushions and pillows run the perimeter of the terrace offering a breathtaking view of the island of Ischia.

Exclusively handcrafted, dark gray stone flooring has been designed and realized by the architect for the entire villa including the baths and kitchen. The architect chose and designed to use the same unique stone for the sinks and showers to maintain the same chromatic flow throughout. The typical Capri Island style is kept through the use of white walls and ceilings but contrasts with the architect's personally designed modern, elegant and minimalistic furnishings. Wood, stone, terracotta and glass become one and add warmth to the villa in complete with the natural surroundings.

POST OPERAM

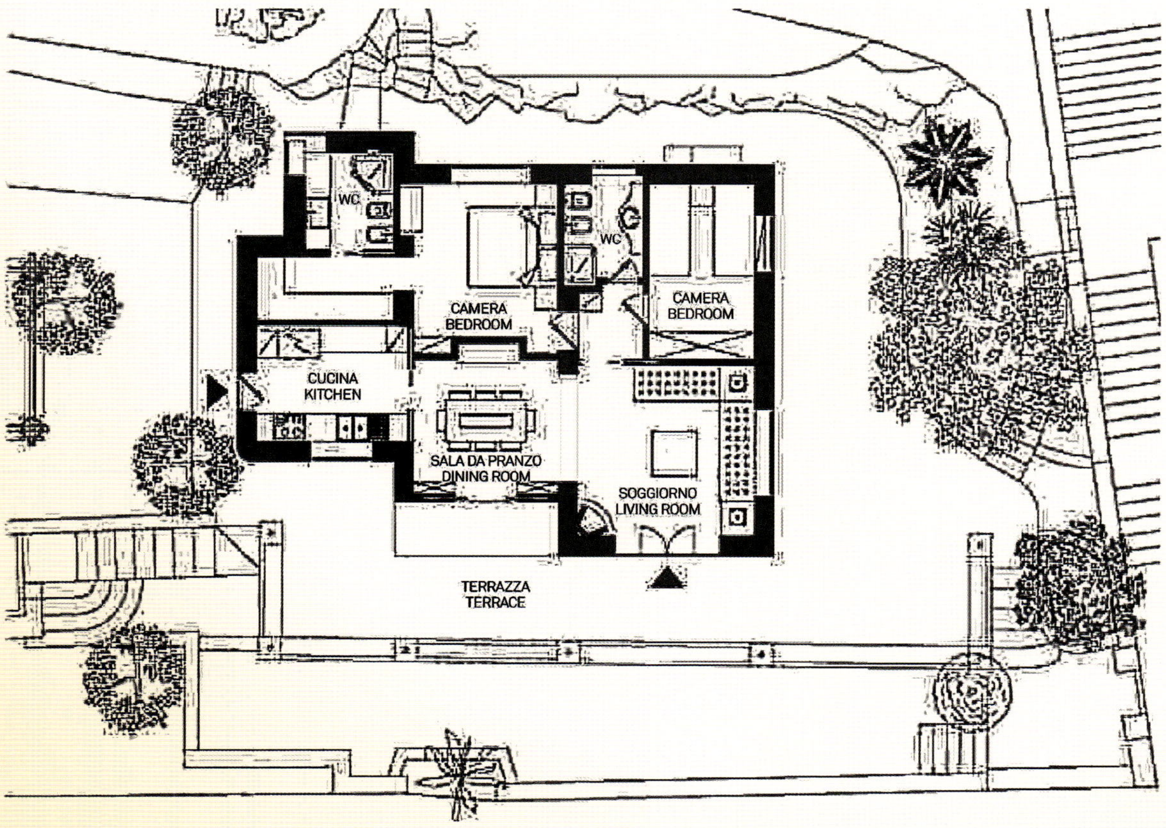
WC
WC
CAMERA
BEDROOM
CAMERA
BEDROOM
CUCINA
KITCHEN
SALA DA PRANZO
DINING ROOM
SOGGIORNO
LIVING ROOM
TERRAZZA
TERRACE

阿纳卡普里别墅在南北两个方向都设有入口。厨房、餐厅、起居室共同形成一个占地 75 平方米的组合空间，通过巨大的玻璃窗可以看到宽敞的户外空间，极大地扩展了视觉空间。从起居室向上，可以看到向外延伸的露台，露台周边安置了弯曲的座椅，上面有坐垫和靠枕，在这里可以欣赏伊斯基亚岛令人窒息的美丽风景。

包括厨房和浴室在内，整个别墅的地面全部由手工打磨的暗灰色石块铺成。为了在整栋别墅中保持色调的流畅和延续，设计师们甚至在下沉地面和浴室里全部采用了相同的暗灰色石块。而使用纯白色的墙壁和顶棚，则体现了典型的卡普里岛风格，与设计师个人选用的现代式家具的优雅与极简抽象风格形成强烈的反差和对比。别墅里还采用了其他装饰材料，如实木、石材、陶瓦、玻璃等等，与周边自然环境共同营造了和谐统一的生活氛围。

The Hotel Hershey

赫尔西酒店

CONTRIBUTOR | *The Hotel Hershey*

LOCATION | *USA*

Tucked in the rolling hills of Central Pennsylvania, you'll find one of America's landmark hotels, The Hotel Hershey, built by chocolate magnate Milton S. Hershey and providing elegant accommodations for guests since 1933.

The Hotel Hershey is located in Hershey, Pennsylvania – known the world over as The Sweetest Place On Earth – a year-round destination with a wide array of attractions and entertainment.

Situated high atop the town of Hershey, The Hotel Hershey offers 276 guest rooms and 2,323 square meters of meeting and function space. It is a recipient of both the Forbes Four-Star Award and the AAA Four-Diamond Award. The Hotel Hershey is also a member of Historic Hotels of America.

An official Resort of Hersheypark, The Hotel Hershey comes complete with The Spa At The Hotel Hershey and access to four distinctive golf courses.

赫尔西酒店是美国的地标性酒店之一，周围群山环绕，最初由巧克力巨头米尔顿·赫尔西出资建造，于1933年正式对外提供住宿服务，环境优雅。

赫尔西酒店位于宾夕法尼亚州中部的赫尔西地区，这里因号称“世界上最甜蜜的地区”而闻名于世，全年都有各种引人入胜的娱乐消遣活动。

酒店地处赫尔西镇的最高处，拥有276间客房以及占地2323平方米的会议区和其他功能区。曾经荣获福布斯四星奖和AAA四钻奖。同时，赫尔西酒店还是美国古老饭店联合会的会员之一。

除了一个官方推荐的度假胜地——赫尔西公园外，赫尔西酒店还设有温泉理疗馆和四个不同的高尔夫场地。

Monte Vista Residence

蒙特 · 维斯塔住宅

ARCHITECT | *Marc Whitman*
FIRM | *Whitman Architectural Design*
LOCATION | *Southern California, USA*
AREA | *1,115 m²*
PHOTOGRAPHER | *Doug A Elis, Micheal Urbanek*

Using an existing ranch style home, this remodel and addition transformed this home into this spectacular Mission Revival Style Estate. The Mission style materials and detailing were borrowed from local historical homes built in the 1920s and 1930s. At the same time new invent detailing were incorporated giving a contemporary flair.

The large outdoor room was built specifically to seat a large wedding party for one of the owner's daughters. Most of the living areas and master bedroom open to this outdoor room which has spectacular views of the Ojai countryside. A large daylight basement on the lower level, with exercise and game room opens to a pool area. The first floor plaster mass anchors the structure to its hillside setting at the same time the rooflines above create a playful composition. Planters and arbors are incorporated to blend the home into its natural setting.

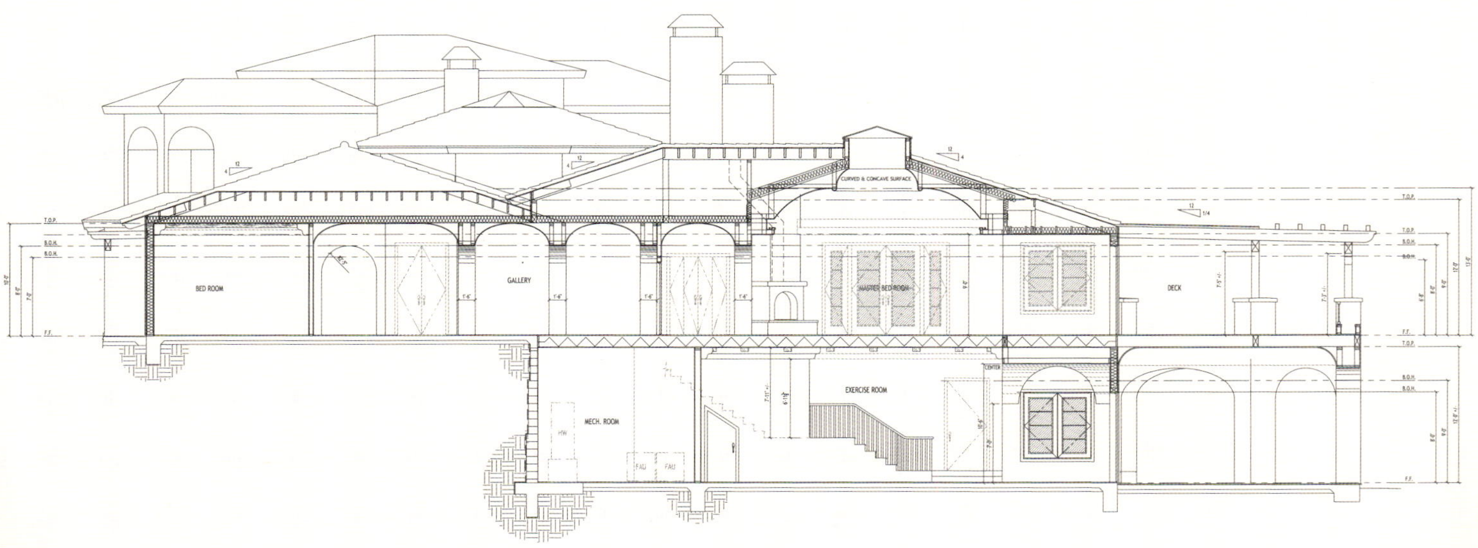
BED ROOM
GALLERY
MASTER BEDROOM
DECK
MECH. ROOM
EXERCISE ROOM

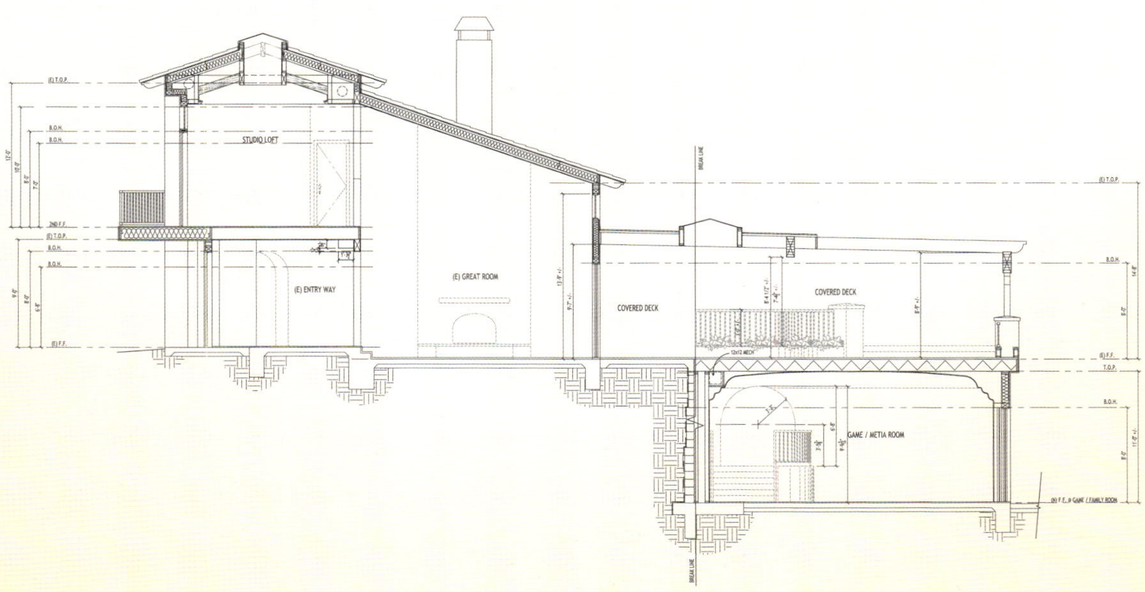
STUDIO LOFT
(E) ENTRY WAY
(E) GREAT ROOM
COVERED DECK
COVERED DECK
GAME / MEDIA ROOM

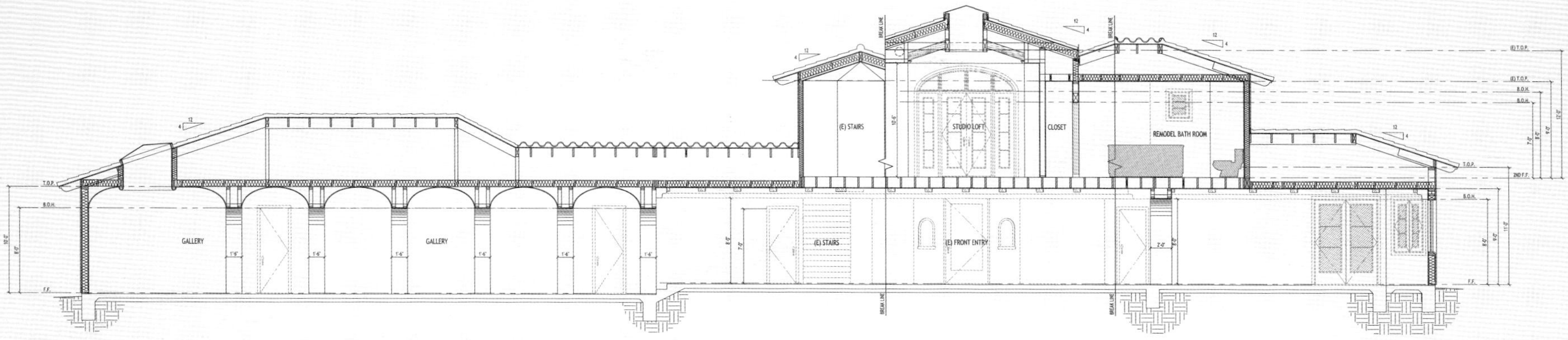
GALLERY
GALLERY
(E) STAIRS
STUDIO LOFT
CLOSET
REMODEL BATH ROOM
(E) STAIRS
(E) FRONT ENTRY

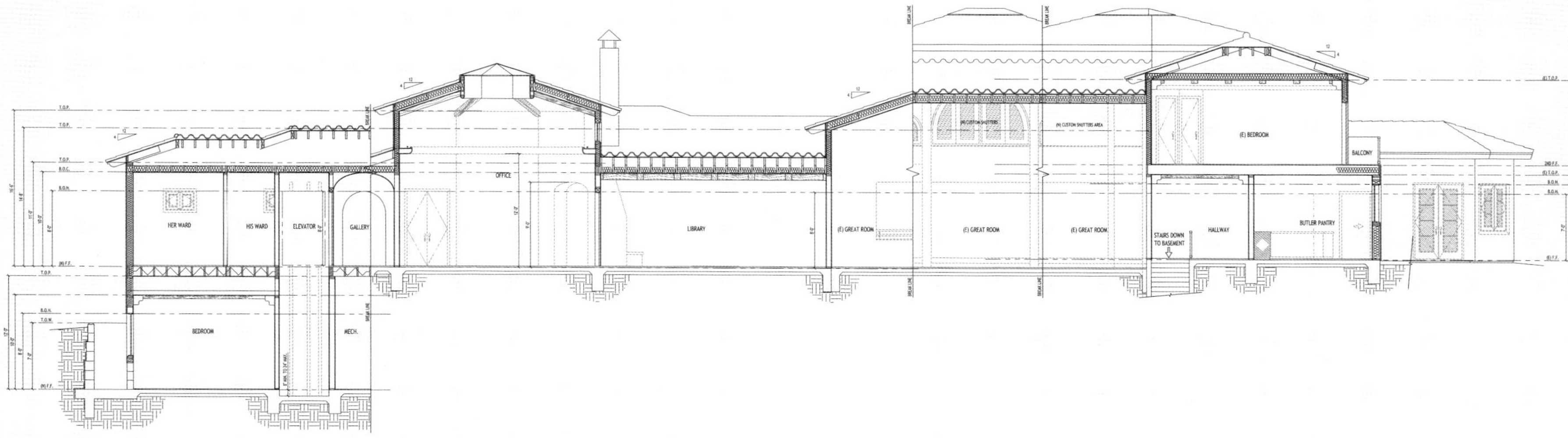
HER WARD
HIS WARD
ELEVATOR
GALLERY
OFFICE
LIBRARY
(E) GREAT ROOM
(E) GREAT ROOM
(E) GREAT ROOM
(E) BEDROOM
BALCONY
STAIRS DOWN TO BASEMENT
HALLWAY
BUTLER PANTRY
BEDROOM
MECH.

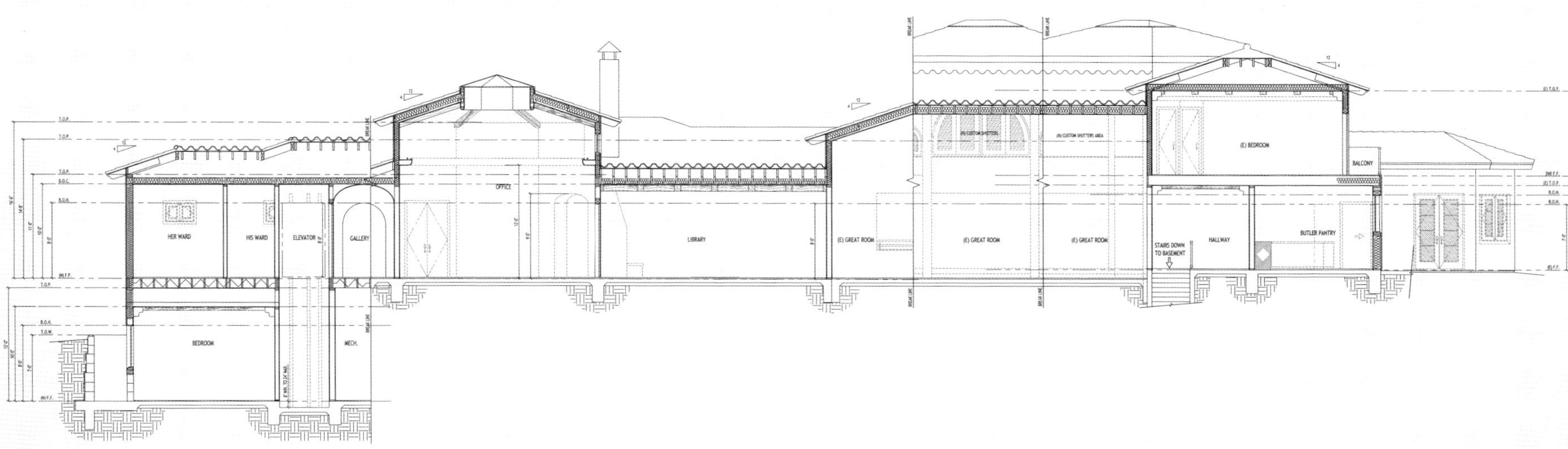
HER WARD
HIS WARD
ELEVATOR
GALLERY
OFFICE
LIBRARY
(E) GREAT ROOM
(E) GREAT ROOM
(E) GREAT ROOM
(E) BEDROOM
BALCONY
STAIRS DOWN TO BASEMENT
HALLWAY
BUTLER PANTRY
BEDROOM
MECH.

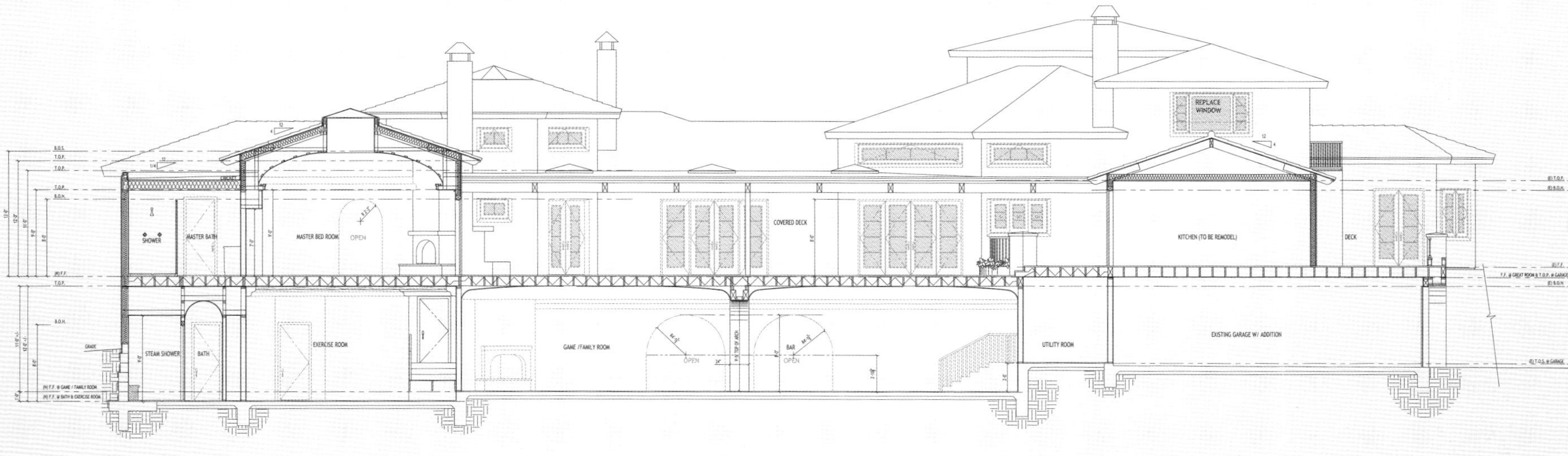
REPLACE WINDOW
SHOWER
MASTER BATH
MASTER BED ROOM
COVERED DECK
KITCHEN (TO BE REMODEL)
DECK
STEAM SHOWER
BATH
EXERCISE ROOM
GAME / FAMILY ROOM
BAR
UTILITY ROOM
EXISTING GARAGE W/ ADDITION

S O U T H E L E V A T I O N

N O R T H E L E V A T I O N

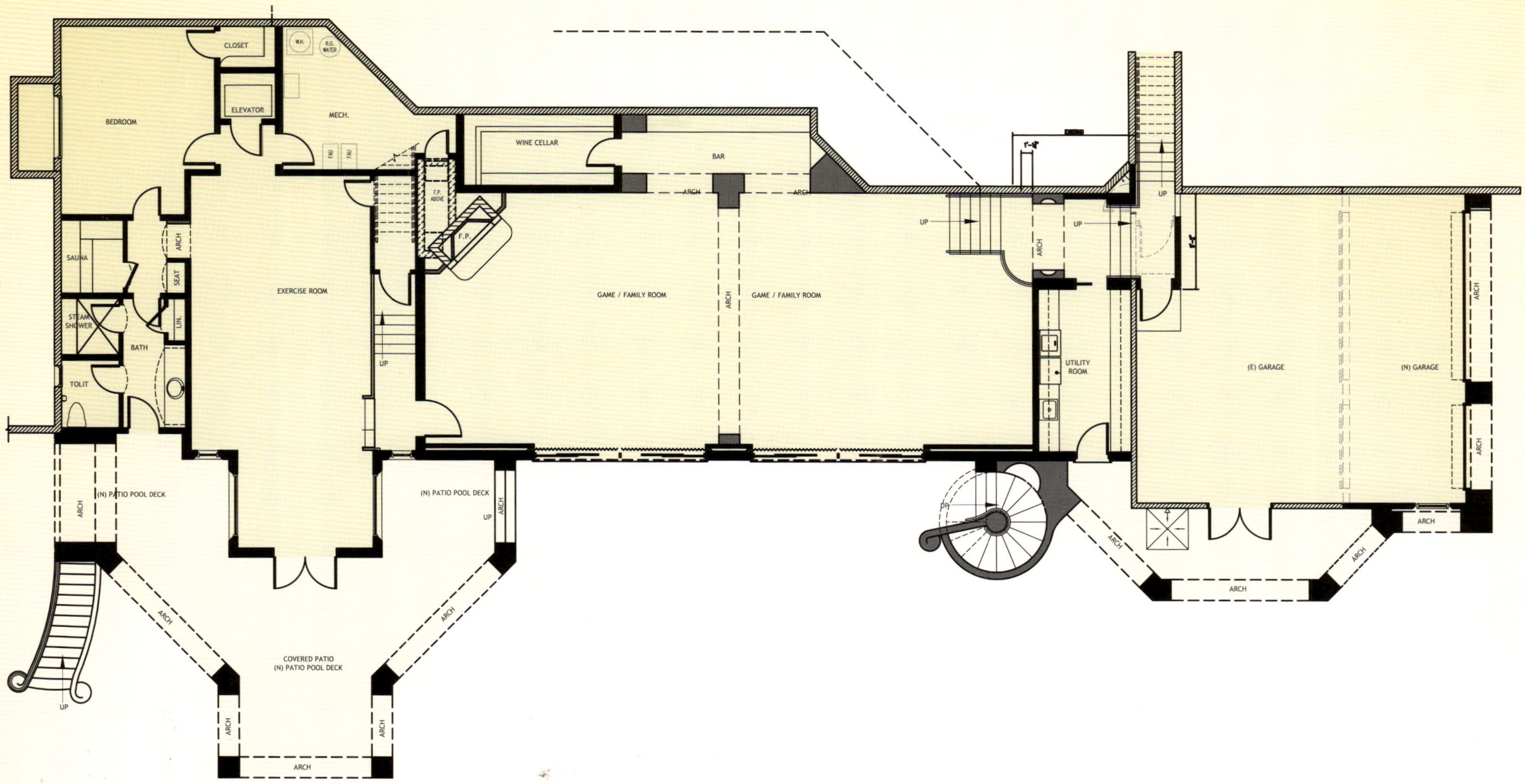

Basement Floor Plan

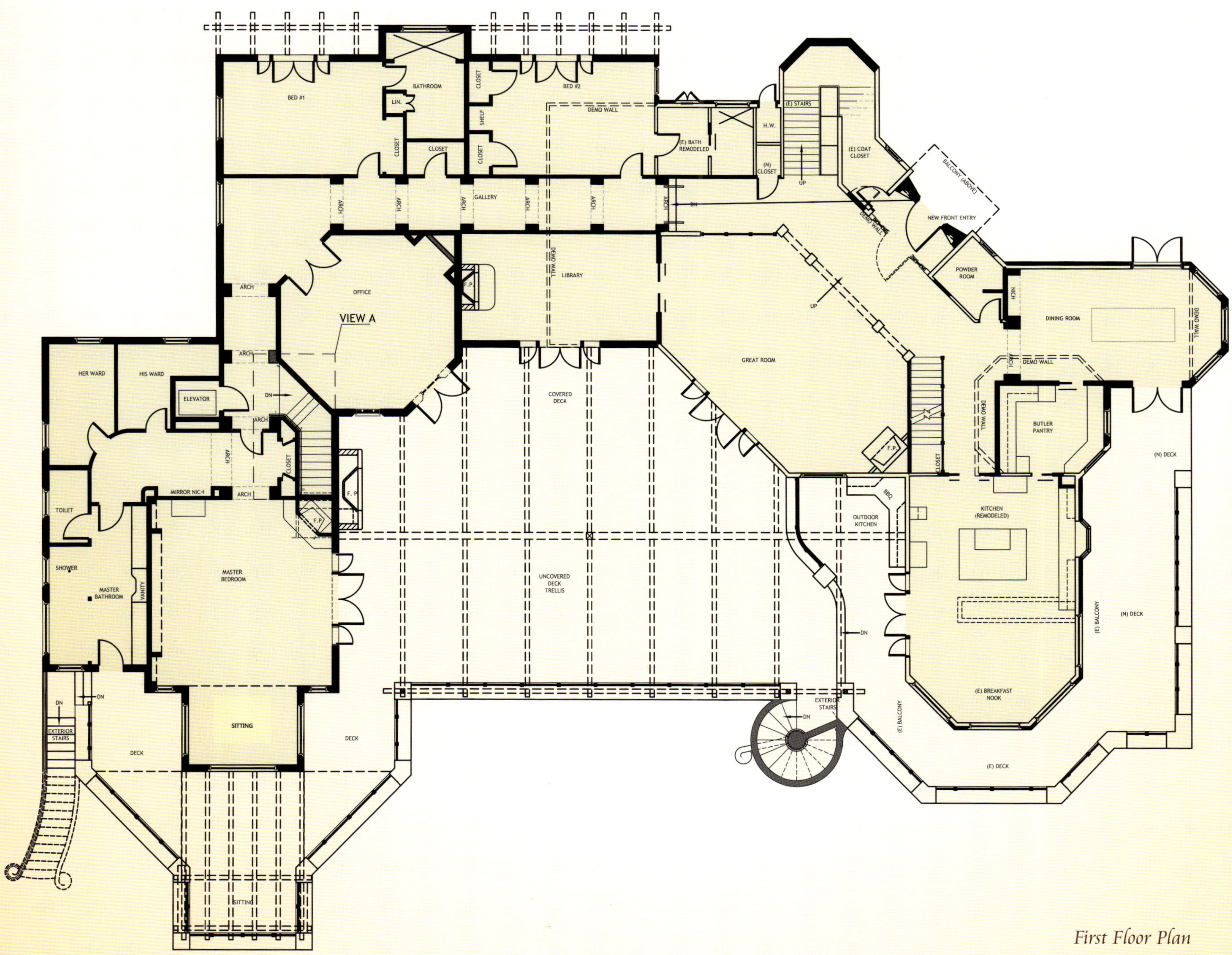

First Floor Plan

蒙特·维斯塔住宅项目是对一个原有的农场风格的住宅进行改造扩建，最终建造出一栋十分壮观的、具有布道院复兴风格的建筑。这种布道院风格和细节设计借鉴了当地建于20世纪二三十年代的历史建筑。同时，设计还融合了一些创新性设计，为整栋建筑增添了些许现代气息。

设计师在室外设计了巨大的活动空间，用来作为房主女儿的婚礼场地。大部分的起居空间和主卧室都朝向婚礼场地，在这里可以欣赏到美丽的奥哈伊乡村的田园风光。低处还有一个宽敞的地下室，里面设有健身房，外面是一个池塘。第一层利用灰泥结构将建筑与后面的山体固定在一起，同时，屋顶的轮廓与山势形成了一个独特的构造。周边大量的花槽和乔木将整个建筑有机地融入到大自然当中。

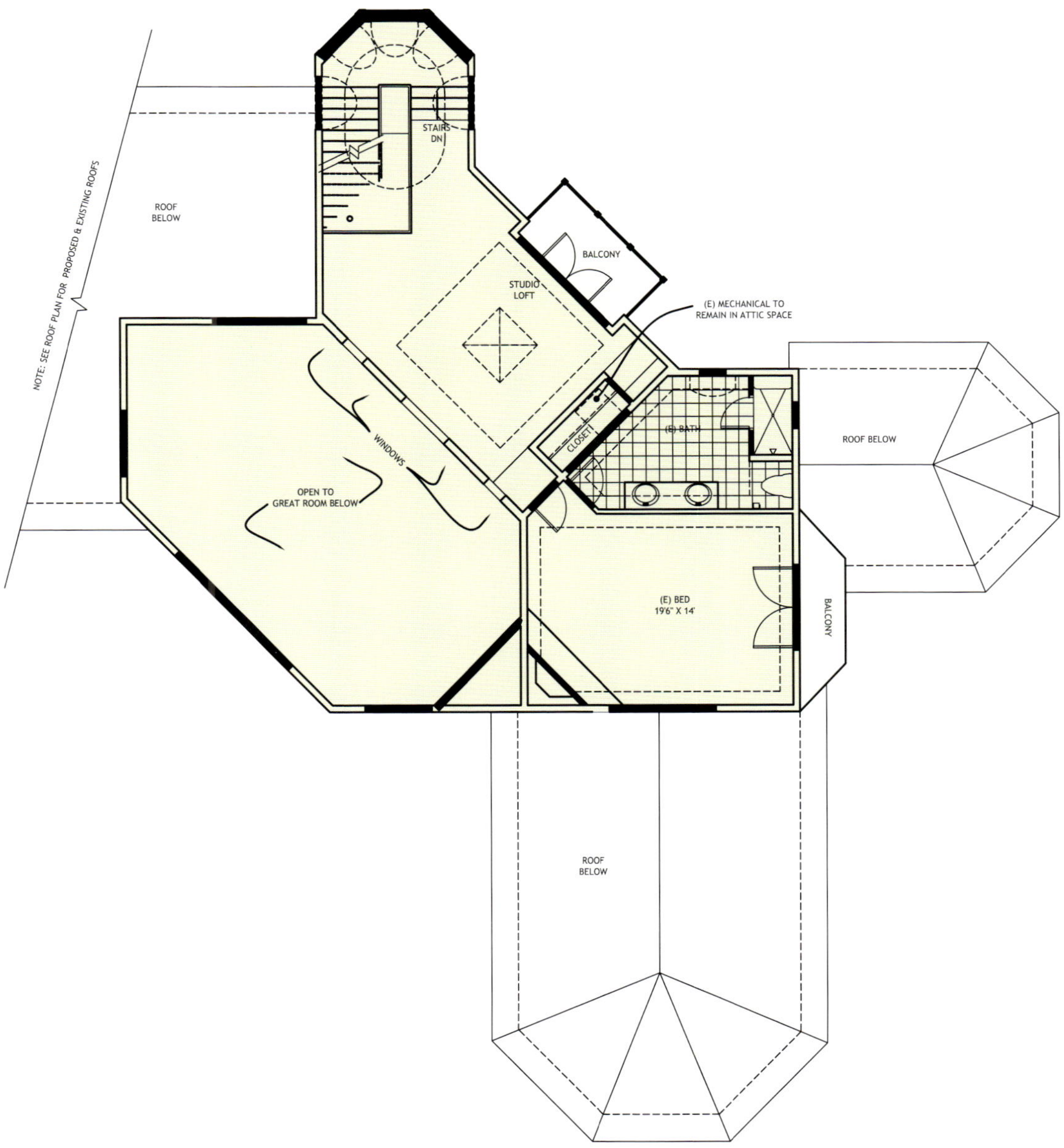

Second Floor Plan

Palisades Beach Road

岩壁下的滨海路住宅

INTERIOR DESIGNER | *Chris Barrett*
DESIGN COMPANY | *Chris Barrett Design*
ARCHITECTURE | *Appleton & Associates Inc.*
CONTRACTOR | *Richard Holz, Inc.*
LOCATION | *Santa Monica, California, USA*
AREA | *414 m^2*
PHOTOGRAPHER | *Alexander Vertikoff*

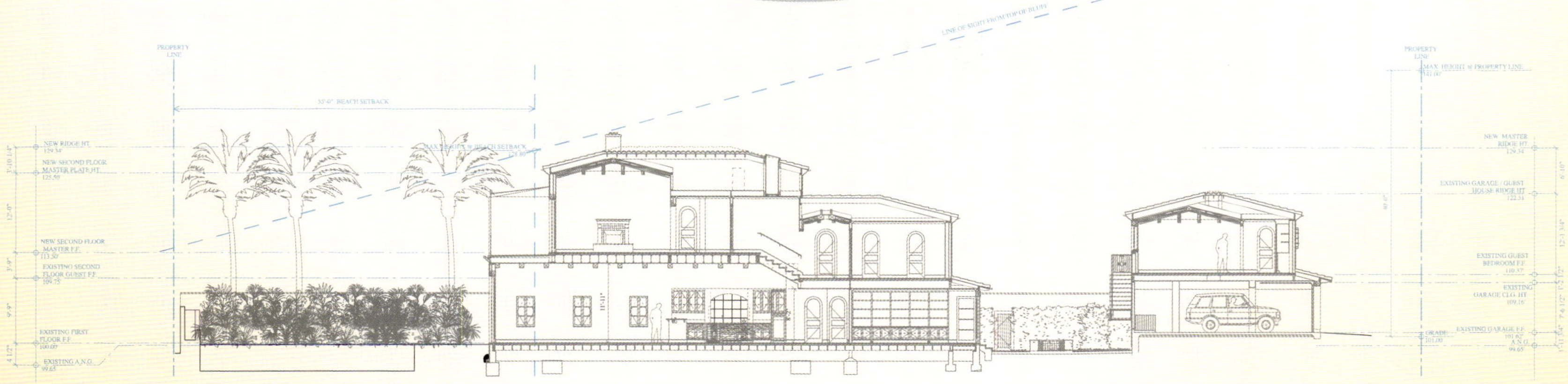

Sitting on the sands of Santa Monica, California, this home responds to its local context but derives its essence from the rich architectural history of the California Spanish colonial and other warm weathered Mediterranean styles. The vision here was to create a home that looks as though it was built in the 1920s and updated for 21st century's living, creating an open loft feeling. Antique beams, white plaster walls and dark wood floors all add to the character of this home.

On the first floor the kitchen, living room and dining room are all open to each other. The high ceilings have antique beams. The walls are white plaster and the floors are a rich dark oak. The kitchen cabinets are dark oak and the counter is a Lagos Azul Caesarstone. The kitchen has everything you need, a glass front refrigerator, a steam oven, microwave, range and a wine refrigerator and more. The living room is a step away and all of the furniture is covered in white linen.

From this area you look beyond the tall arched French glass doors, which lead you to the pool in the courtyard, covered with antique French Terra Cotta pavers. The plaster fireplace was custom designed, as was all of the tile work at the pool area. Tucked behind the staircase is a powder room and a media room with handsome wood ceilings which open on to a charming courtyard and fountain. This courtyard is flanked on the opposite side with the garage and a staircase leading up to the carriage house (guest quarters).

The second floor has a laundry room with antique Spanish cement tiles, a guest bedroom and bath and the master suite. The master bath has a custom vanity and dome ceiling and the shower is tiled with 5.08cm×5.08cm white glazed Moroccan terra cotta tile. At the master tub, there is a sliding wood screen that opens the bathroom to the bedroom and beyond.

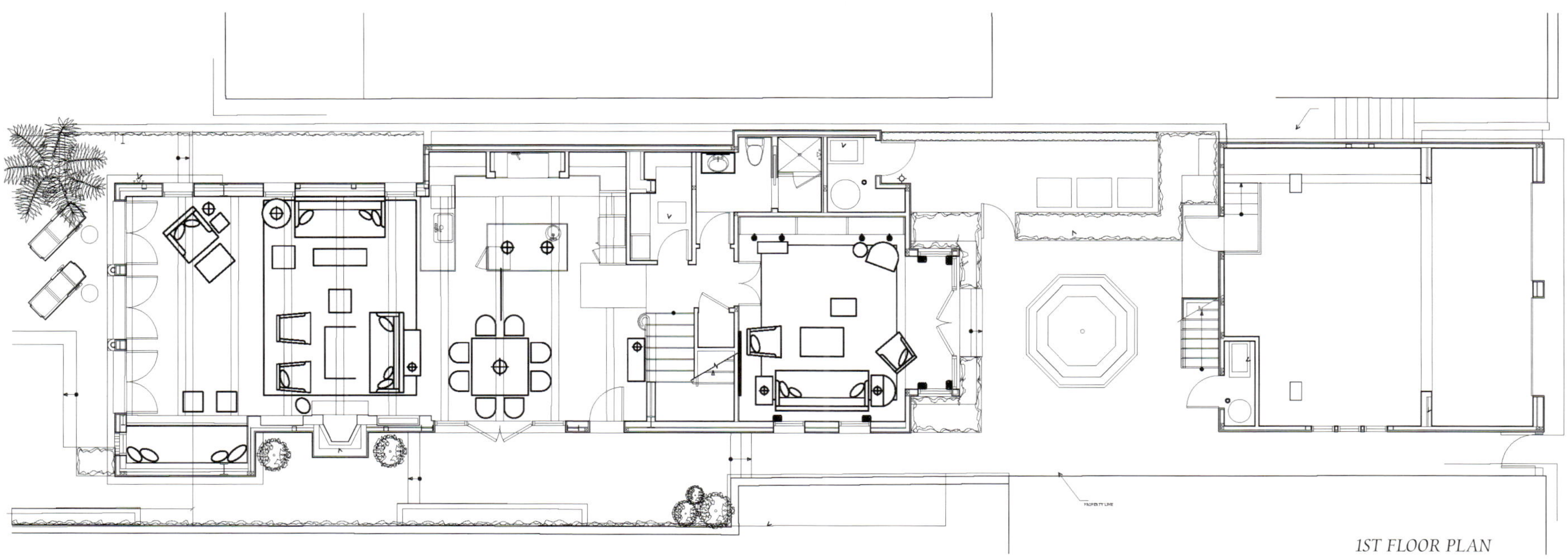

1ST FLOOR PLAN

Black Star

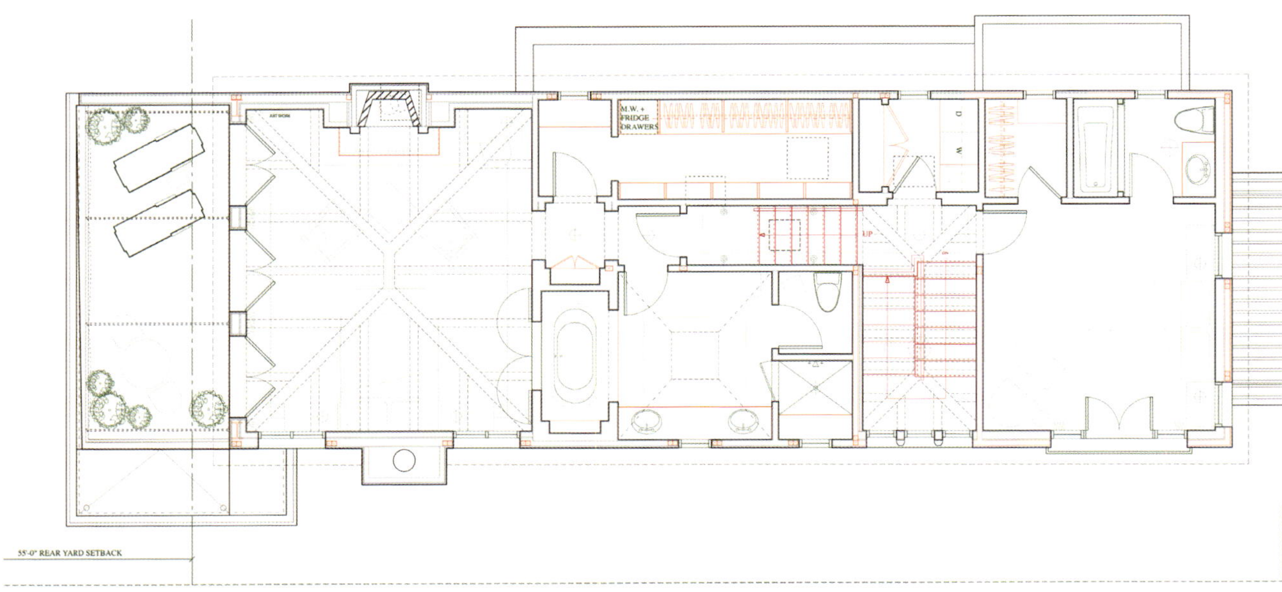

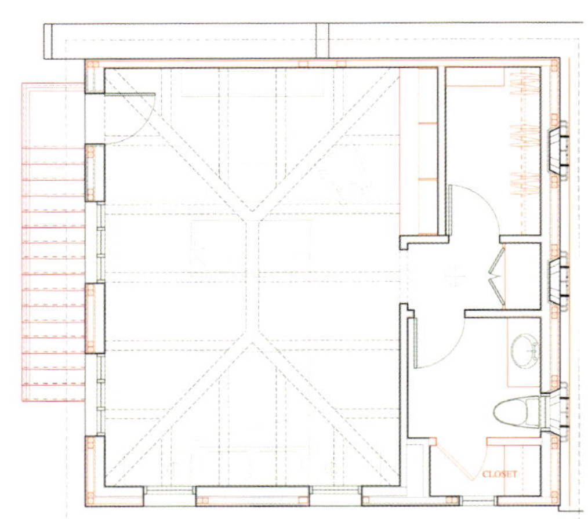

2ND FLOOR PLAN

这栋住宅位于加利福尼亚圣莫妮卡海滩边，建筑构造具有典型的当地建筑特点，混合了加利福尼亚历史悠久的西班牙殖民主义以及温馨舒适的地中海建筑风格。改建的目的是将这栋看上去极具 20 世纪 20 年代风格的建筑进行改造升级，加入一些适用于 21 世纪的元素，打造一个开放式的建筑模式。仿古的房梁、白色粉泥墙体和暗色实木地板，都充分体现了十分鲜明的建筑特色。

地面一层是厨房、起居室和餐厅，都是开放式的结构。高高的天花板上交错着仿古式的房梁，墙体是白色粉饰泥灰，地板是黑色橡木地板。橱柜也是黑色橡木制成，台面是 Lagos Azul 恺萨金石。厨房里设施齐备，带有玻璃面板的冰箱、蒸汽炉、微波炉、冷藏酒柜等等。只有一步之遥的起居室里所有的家具都是白色的，从这里可以直接通到高高的法式玻璃拱门，看到外面庭院里的池塘，地上铺满了法兰西赤土陶色地块。灰泥壁炉和池塘边的瓷砖都是运用传统方式铺就而成。楼梯拐角处是一间化妆室和媒体室，上面是木质天花板，窗外是庭院和喷泉。庭院的侧面是停车场，旁边有一个楼梯通往客房。

二层设有一个布满了仿古西班牙式瓷砖装饰的洗衣房、客卧、浴室和主套房。主浴室充斥着传统的奢华，高高的穹顶，四周使用 5.08 厘米见方的亮白色摩洛哥式套图瓷砖砌成。主浴缸旁边摆放了一个木质滑动屏风，将浴室和卧室完美结合到一起。

State and Hope

国家和希望公寓

ARCHITECT | Peikert Group Architects, LLP

This mixed-use project consists of 6 luxury condominiums over a parking structure and 465 square meters of commercial space at street level in the upper State street area of Santa Barbara.

这个多用途的托管公寓坐落在圣巴巴拉北部的国家大道区域，设计包括 6 个奢华的部分，外加一个停车场和一个 465 平方米的商业区。

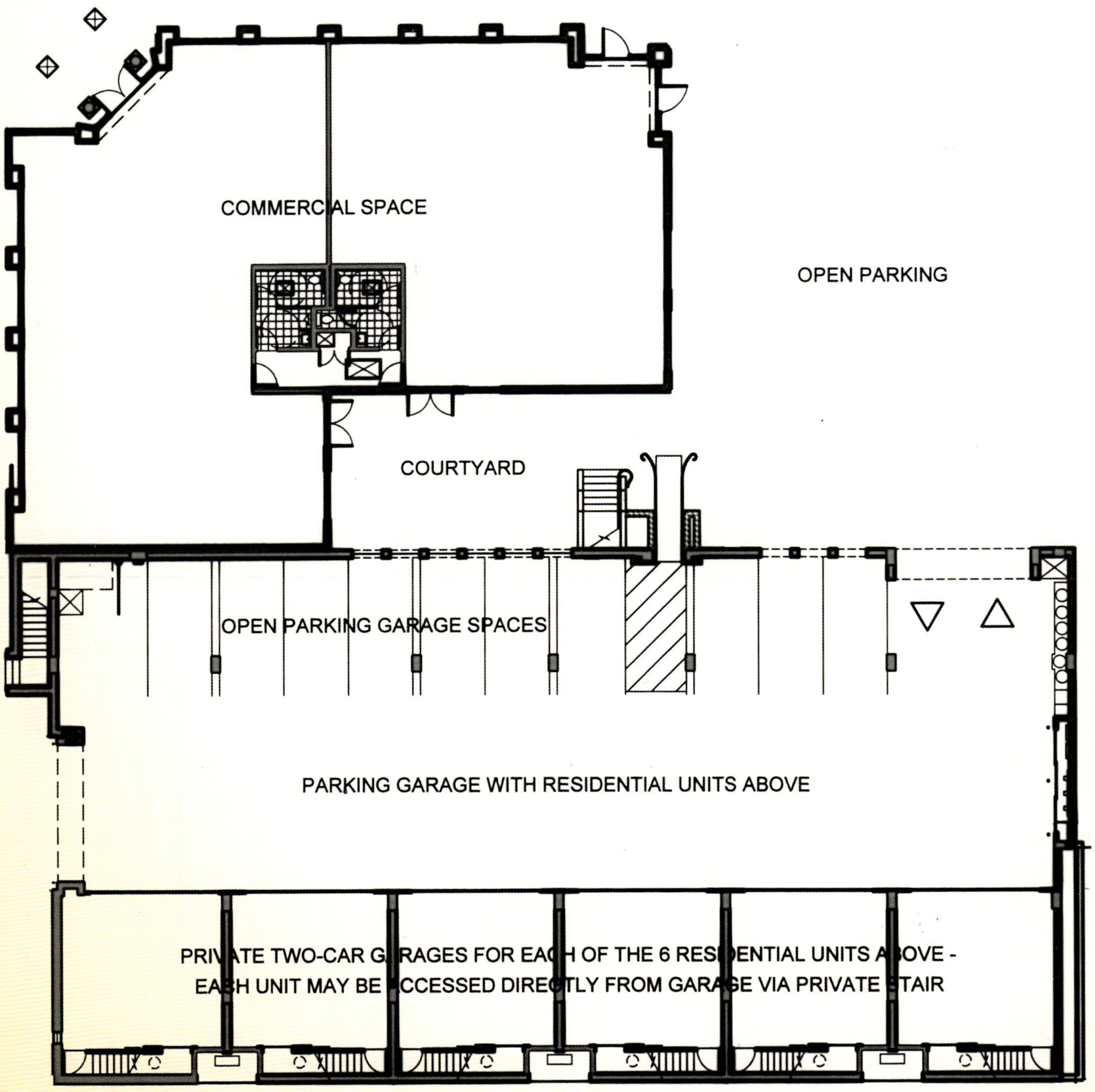

FIRST FLOOR PLAN

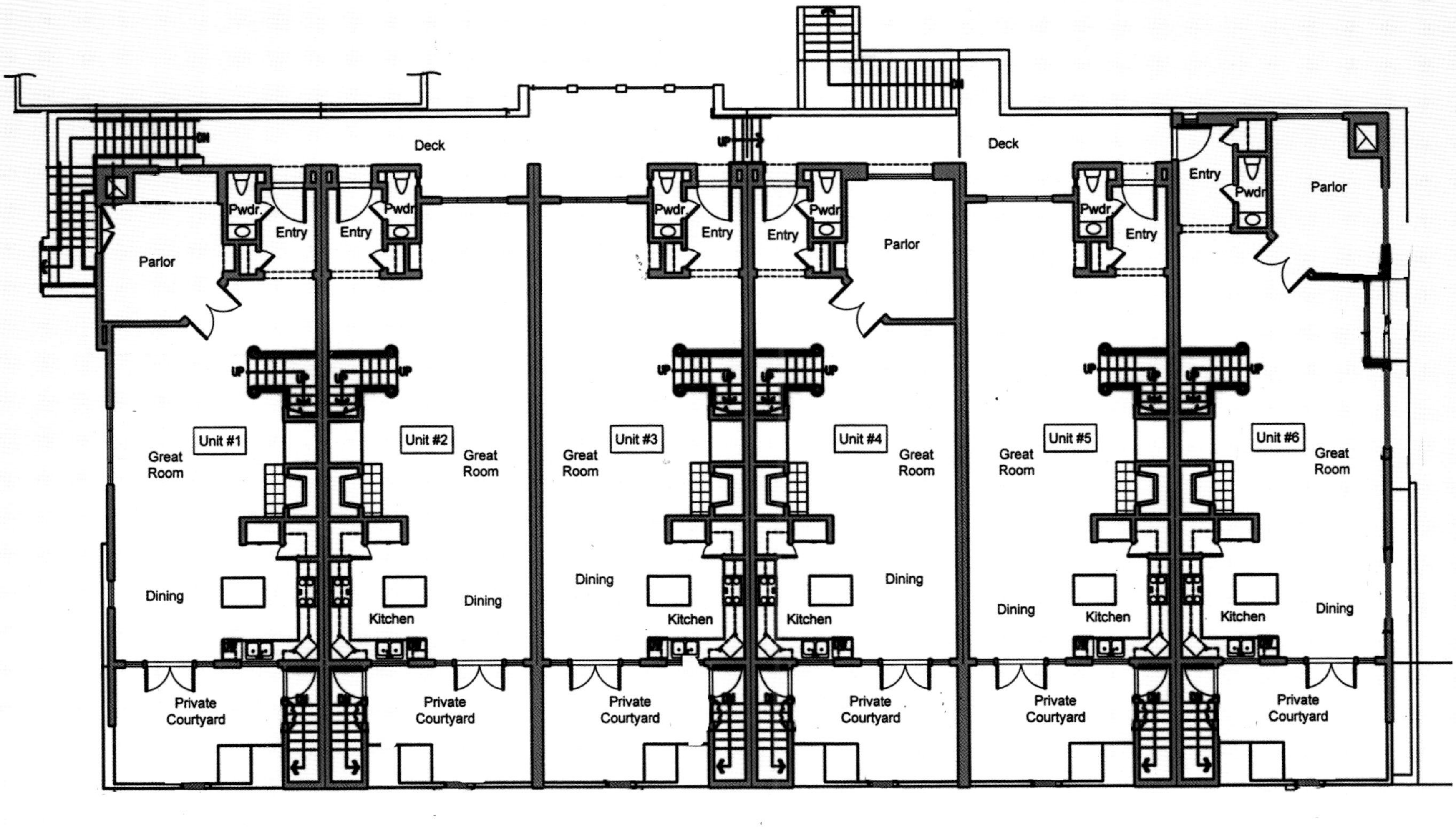

SECOND FLOOR PLAN

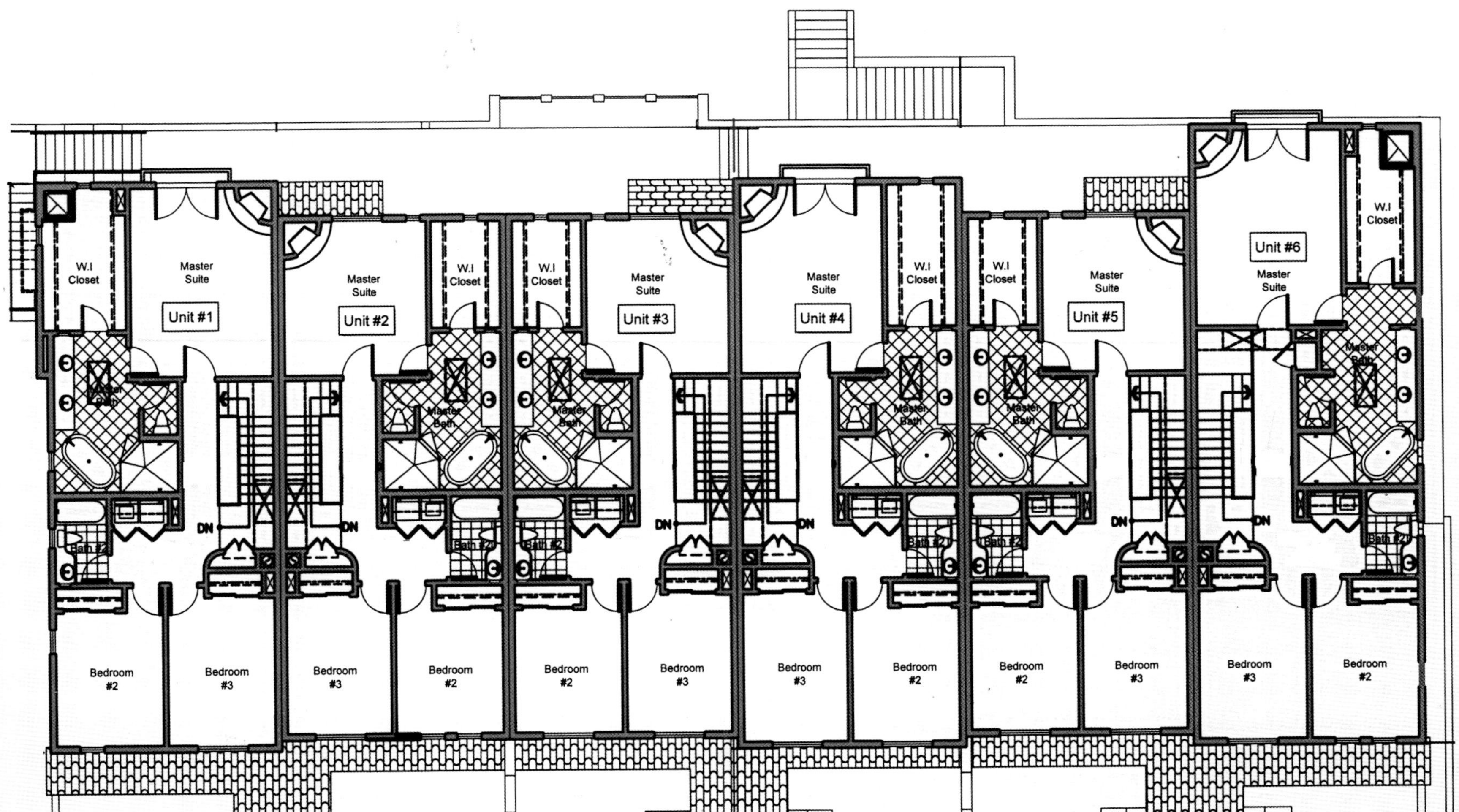

THIRD FLOOR PLAN

Spanish Oaks

西班牙栎林住宅

ARCHITECT | Hugh Jefferson Randolph Architects
PHOTOGRAPHER | Trey Hunter

This home, in the west of Austin Texas, is one of a neighboring pair of homes designed to be in the spirit of the 1920s Spanish Revival style that was most common in California cities such as Santa Barbara and Los Angeles. Set in a neighborhood with design guidelines, key goals were simplicity of form, utilization of the natural slope, and the highlighting of views. Rather than following the lead of the neighboring houses, these homes were designed to be enclaves unto themselves, both physically and stylistically.

Special focus was given to the contrast between the simple clean white stucco walls and the more textured and crafted accents. Materials such as antique roof tile, salvage wood beams, artistic glazed floor tiles and hand carved stone elements were used to provide visual delight and authenticity.

NORTH ELEVATION

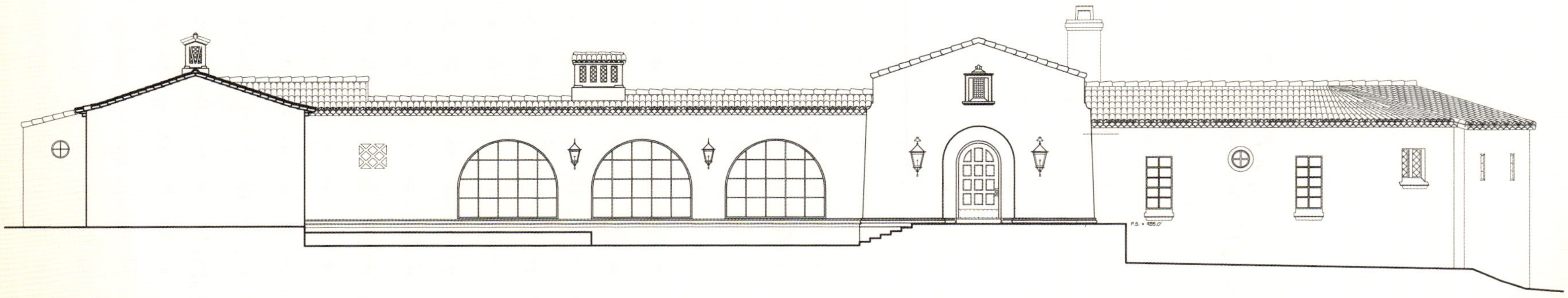

SOUTH ELEVATION

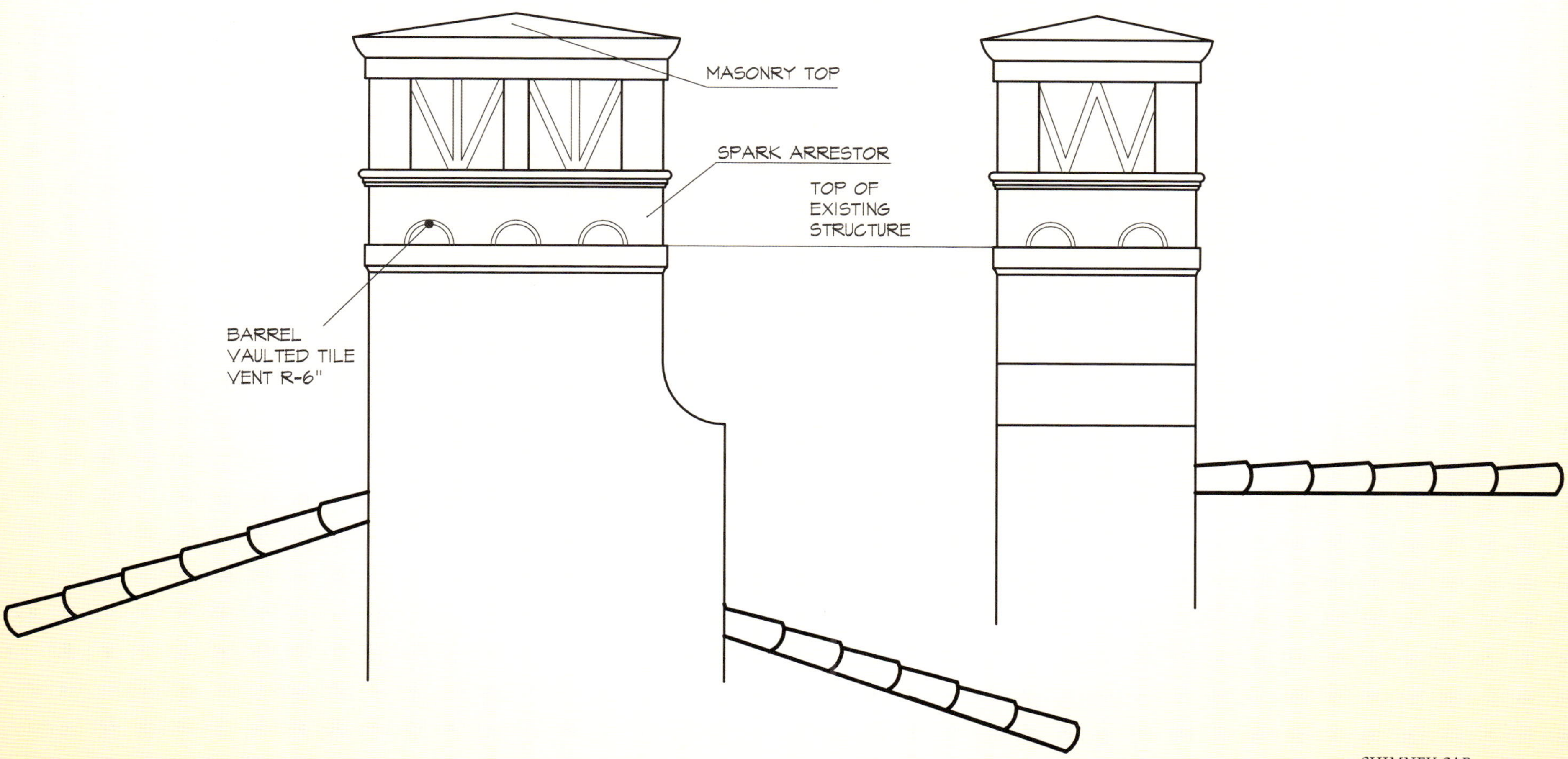
MASONRY TOP
SPARK ARRESTOR
TOP OF
EXISTING
STRUCTURE
BARREL
VAULTED TILE
VENT R-6"

CHIMNEY CAP

MAIN FLOOR

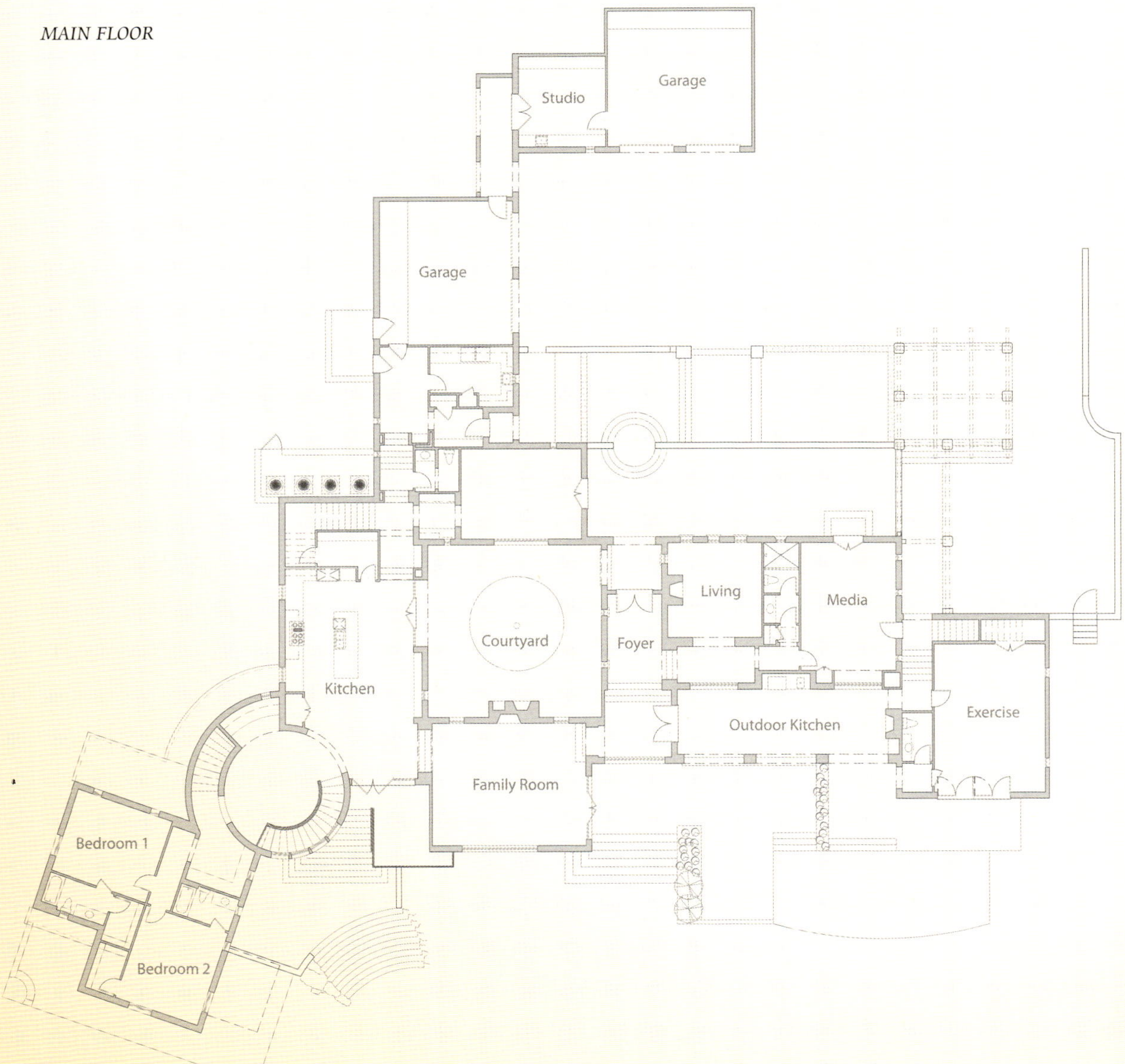

FRONT ENTRY

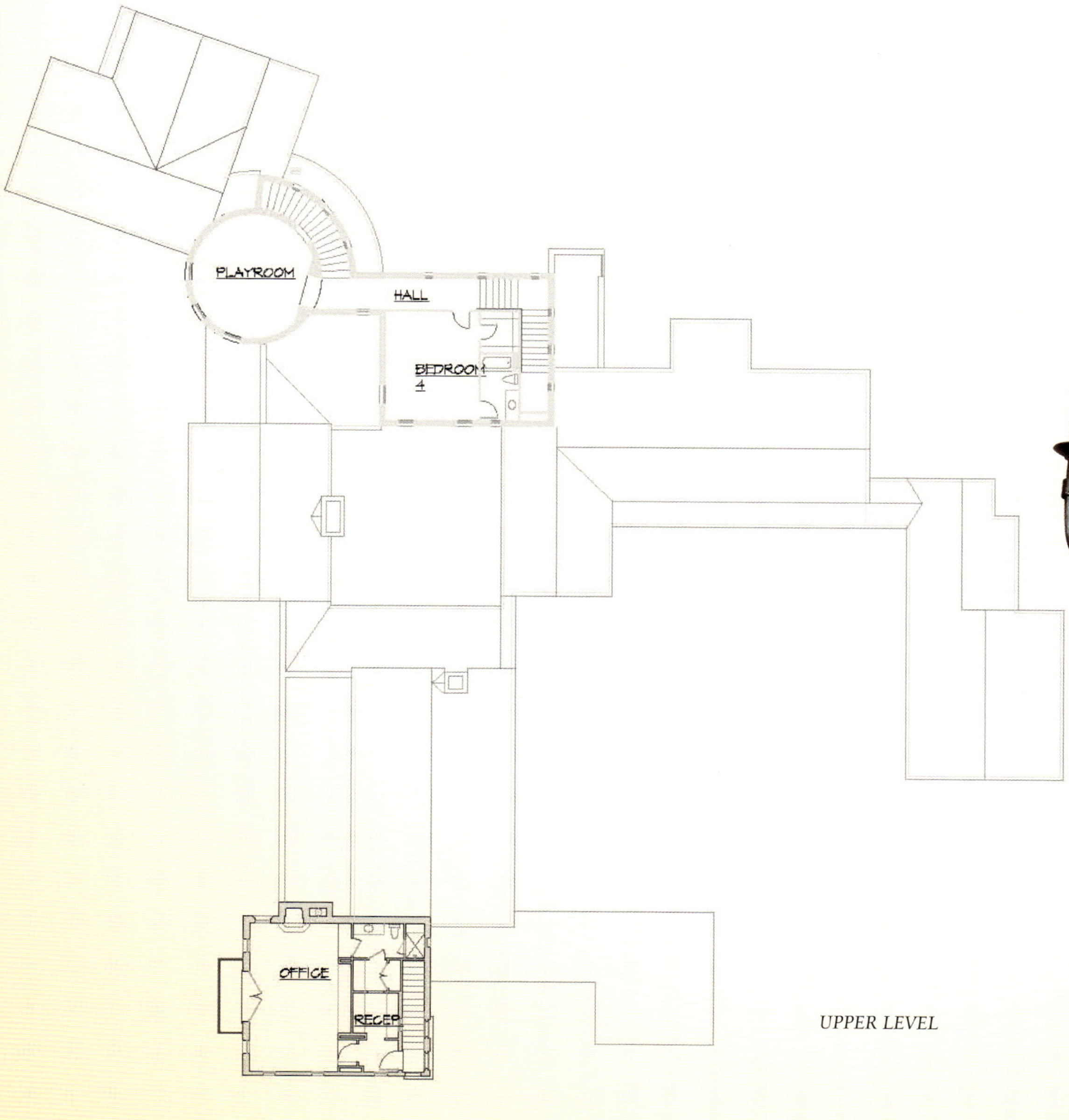

UPPER LEVEL

这栋住宅位于德克萨斯州奥斯丁西部，是一对典型的20世纪20年代西班牙复兴时期风格的住宅中的一栋，这种风格在加利福尼亚州的洛杉矶和圣巴巴拉地区十分常见。设计师们选取了与周边建筑不同的设计理念，主要目的是充分利用自然的斜坡元素，设计营造出外形质朴的建筑外观，充分突出周边的美丽景色。与相邻房屋千篇一律的设计完全不同，住宅中的房屋在设计上都独树一帜，无论是房屋结构本身还是房屋的设计风格，每一处房屋都是一处属于自己的天地。

设计师们十分关注简单整洁的白色粉饰灰泥墙体与手工制作的具有粗糙织纹的装饰之间的反差。同时，使用了不同的建筑材料，如仿古式的屋瓦、回收重新利用的实木房梁、艺术气息强烈的釉面地砖以及手工雕刻的各种石质装饰，为整个建筑提供了一种极具真实感和愉悦感的视觉效果。

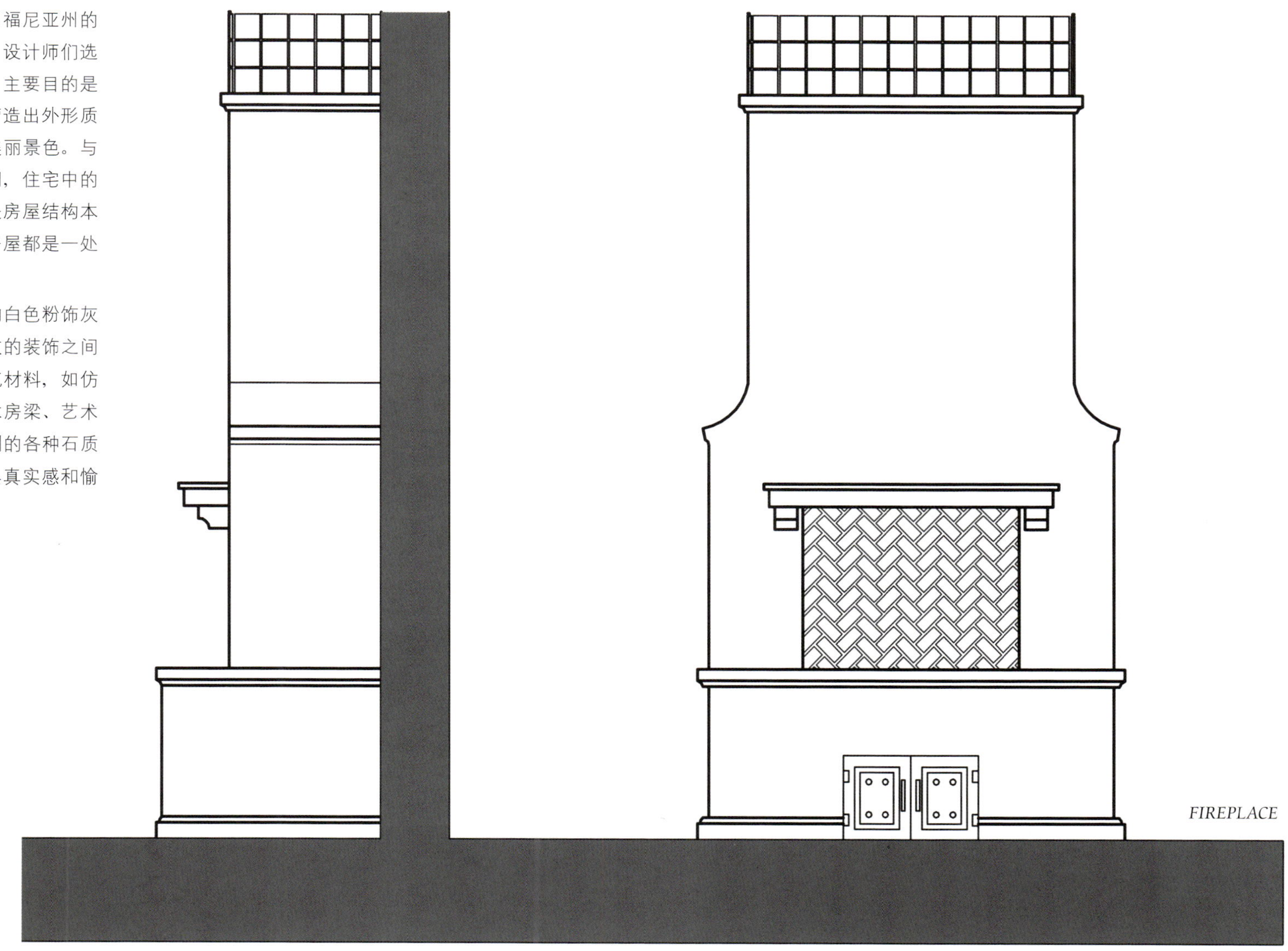

FIREPLACE

Casa Lluvia Blanca

路维亚·布兰卡住宅

ARCHITECT | *Cathi House*
FIRM | *House + House Architects*
LOCATION | *San Miguel de Allende, Mexico*
AREA | *418 m^2*
PHOTOGRAPHER | *Steven & Cathi House*

Nestled into the quiet middle of a block in the historic center of the beautiful colonial town of San Miguel de Allende, this 418 square meters courtyard home is accessed through lush gardens with trickling fountains and a luminous lap-pool. The living room, dining room, kitchen, library and master suite on the first floor open onto a series of plant filled patios that flood each space with light that changes throughout the day. Elliptical domes and hewn wooden beams sculpt the ceilings, reflecting soft colors onto curving walls. A long, narrow stairway wrapped with windows and skylights is a serene connection to the second floor "Moroccan" inspired suite with domed fireplace and hand-sculpted tub, and "French Country" inspired suite with a sunny balcony and oval shower. A curving bridge flies through the high living room with sparkling glass railings and overlooks onto sensuously shaped built in sofas. At the third floor windows wrap every space with balconies, light and views, linking indoors to the distant mountains, the morning sun and the bubbling jacuzzi. At the rooftop terrace domes and chimneys join the cozy seating for intimate gatherings.

Section A cut through rear garden, bedroom 1, central courtyard, window seat at living, front courtyard, bedroom 2, sunset terrace & studio - looking north

Section B cut through bath 1, central courtyard, dining, kitchen, laundry, bedroom 3, bath 3 & studio - looking north

Section C cut through entry courtyard, living room & library - looking south

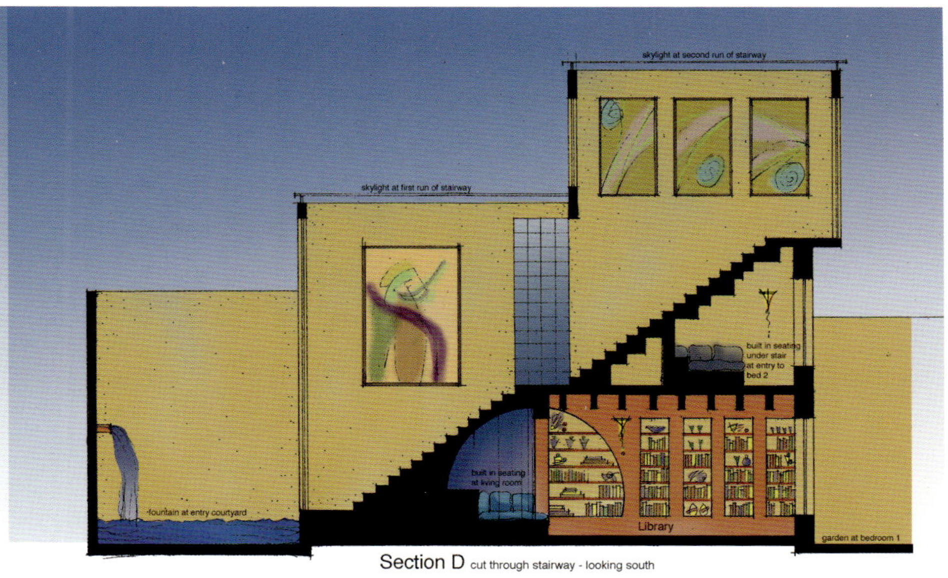

Section D cut through stairway - looking south

Living Room looking north to dining

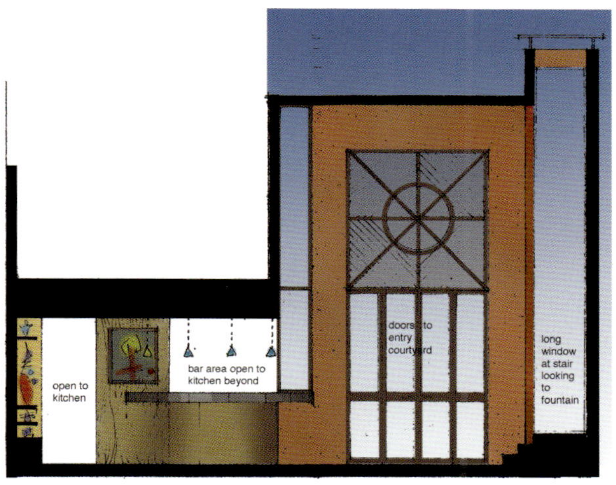

Dining looking east to kitchen Living Room looking east to entry courtyard

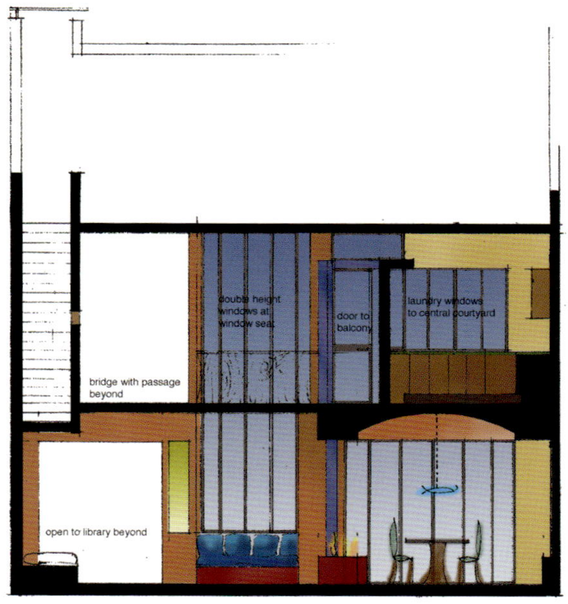

Living looking west to library Dining looking to central courtyard

Bedroom 1 looking east to central courtyard

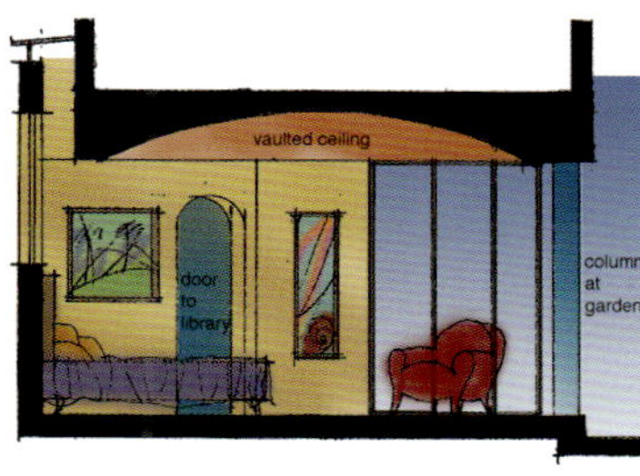

Bedroom 1 looking south to town

Bedroom 1 & Dressing looking west to private garden

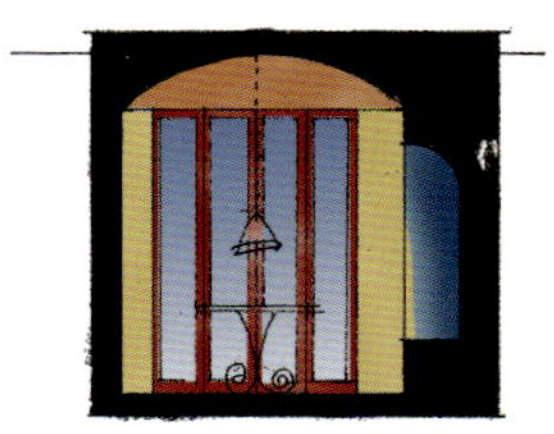
Nook looking east to pool

Kitchen looking east to nook

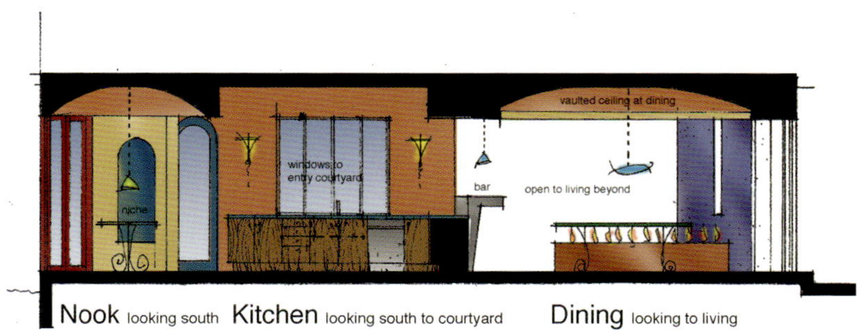

Nook looking south Kitchen looking south to courtyard Dining looking to living

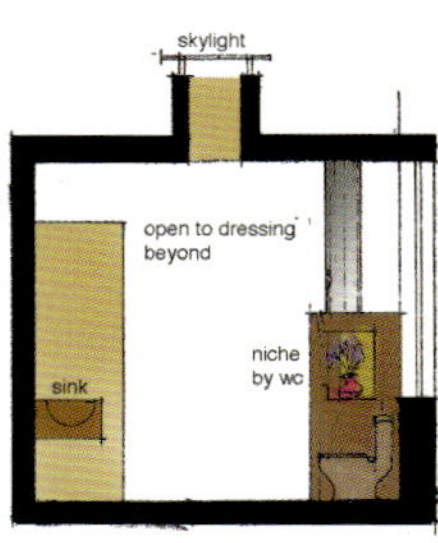

Bath 1 looking to dressing

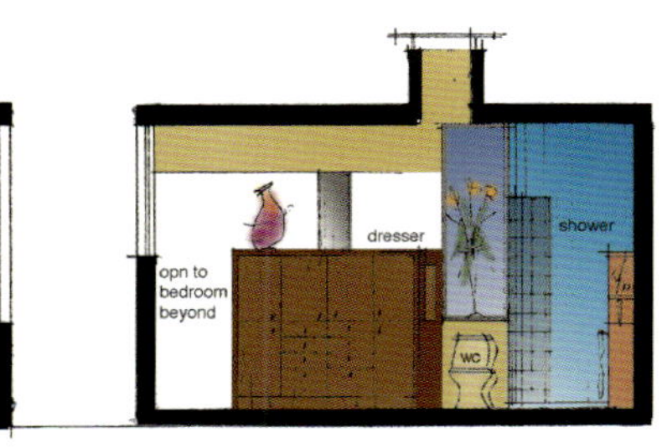

Dressing Bath 1 looking south

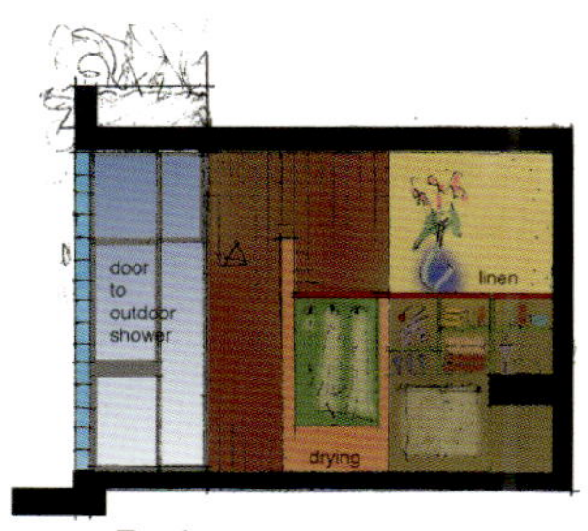

Bath 1 looking west

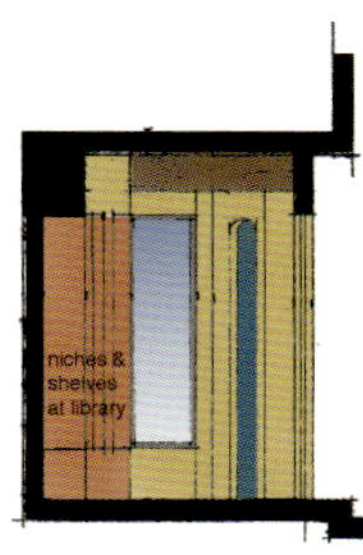

Library looking west

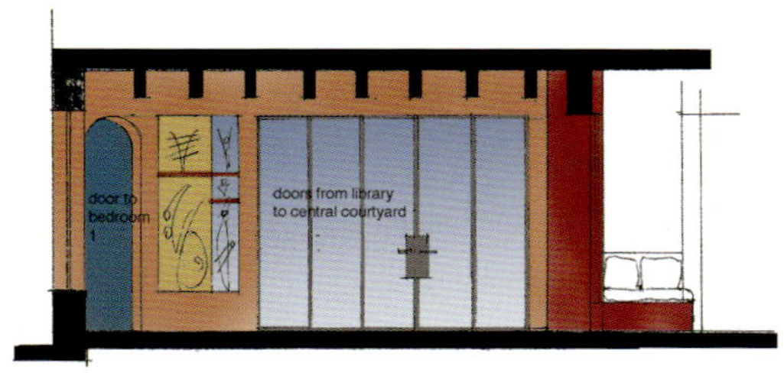

Library looking north to central courtyard

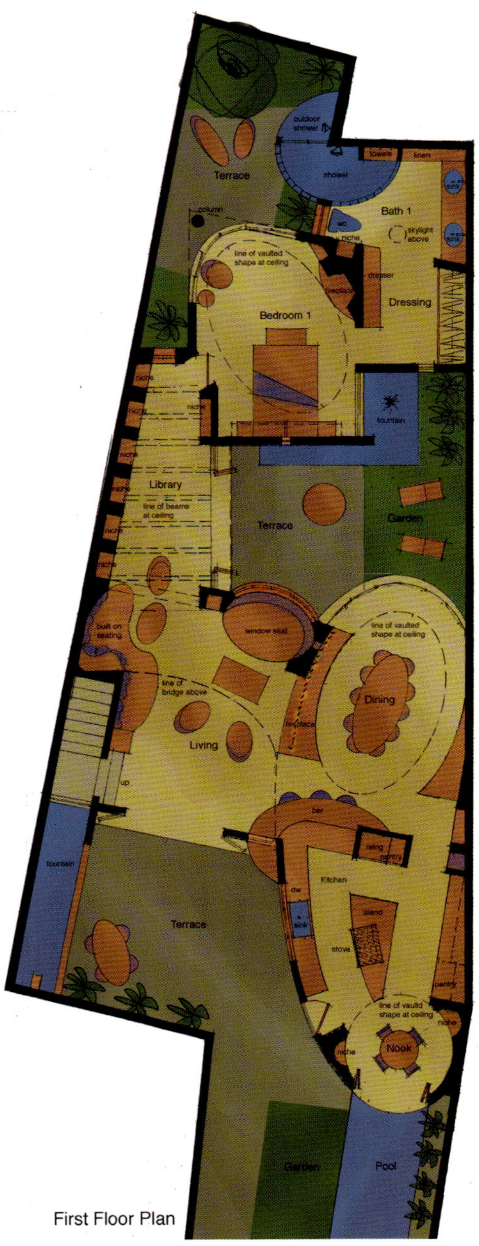

First Floor Plan

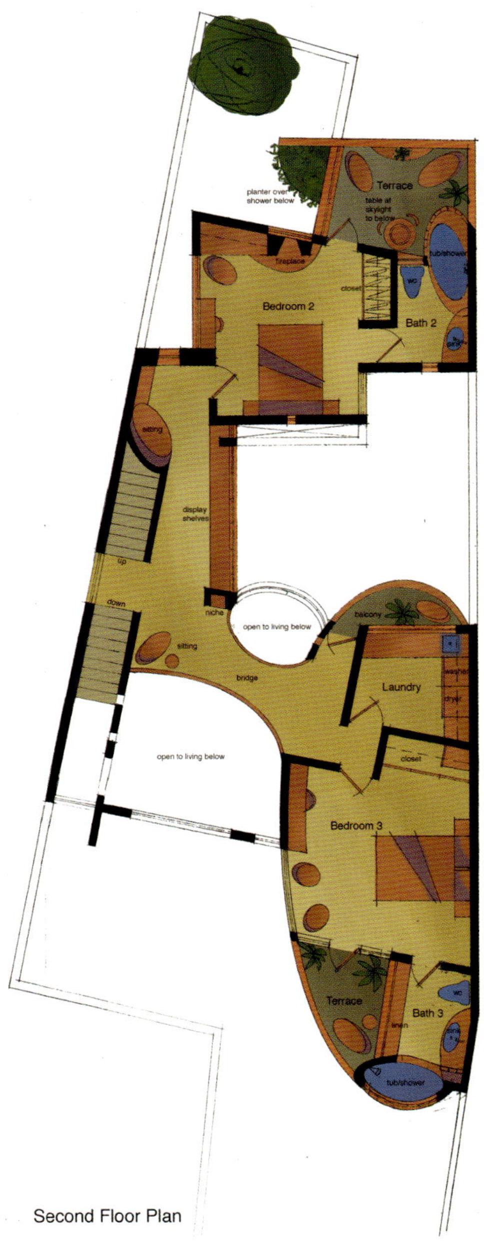

Second Floor Plan

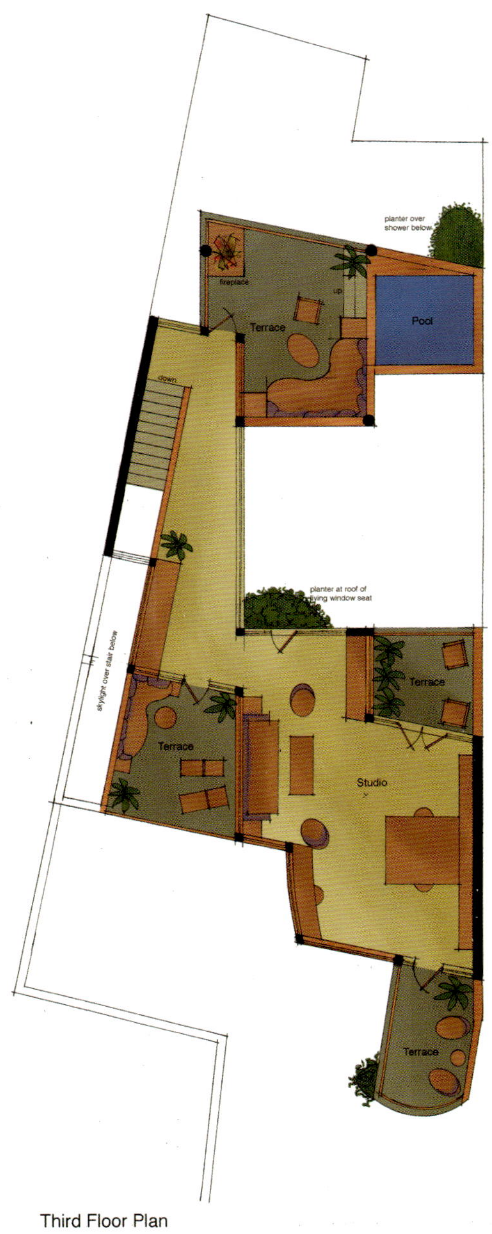

Third Floor Plan

路维亚·布兰卡住宅位于圣明戈迪艾兰德幽静的街区中，仿佛置身于历史长河的中心，这里是一处美丽的殖民地风情小镇。住宅庭院占地面积约418平方米，可直通郁郁葱葱的花园，花园中有汩汩的喷泉和波光粼粼的小型健身泳池。起居室、餐厅、厨房、藏书室和主人的套房都设在一层。穿过一层可以直接进入种满各种植被的院子，独特的设计保证了这里的每一个角落在白天都有充足的光照。椭圆形的屋顶和雕刻的木梁装饰的天花板将柔和的光线反射到弧形墙壁上。一处狭长的楼梯盘旋而上，掠过窗户和天窗直到二层，这种衔接显得自然而流畅。二层设有圆顶壁炉和手工雕刻的浴盆，每一个细节都彰显着摩洛哥风情。洒满阳光的阳台和椭圆形的淋浴间却透露着法国乡村的浪漫。奢华的起居室设计了拱形的窗梁和带栏杆扶手的玻璃落地窗，坐在沙发里就有一种可以俯瞰整个景观的感受。在三层，窗外的阳台、光线和景观三者将室内装饰与外在环境融为一体，躺在极可意泡泡浴缸里即可观赏远处的群山和日出的太阳。屋顶拱形棚顶的露台上，闲暇时光可邀三五好友，在圆形灯罩下，享受惬意的美好时光。

Norton Avenue Residence

诺顿大街住宅

DESIGNER | Karen Vidal, Guy Vidal
DESIGN COMPANY | Design Vidal
LOCATION | Glendale, Los Angeles, Ca, USA
PHOTOGRAPHER | Curtis Dahl

Karen Vidal was hired by the owners, a young couple, to help co-design their dream house in the neighborhood of Glendale, Los Angeles. Karen was given charge of the living and dining rooms in this Moroccan style house. All furniture, cabinetry and built ins, as well as the parking gate, were custom designed by Design Vidal and custom built for this home.

851

这栋住宅位于洛杉矶附近的格伦达尔。业主是一对年轻的夫妇，他们聘请了卡伦·维达尔作为住宅建筑的合作设计师，为他们设计建造一栋理想的住宅。住宅的整体设计为摩洛哥风格，卡伦负责起居室和餐厅部分。所有的设施、细木家具、内部装饰乃至于停车场的大门，都是按照业主的要求，由维达尔设计公司定制完成的。

BIBI JORDAN
SWAHILI CHIC
THE FENG SHUI OF AFRICA
THE
1000
BEST RECIPES
bay books

DALÍ

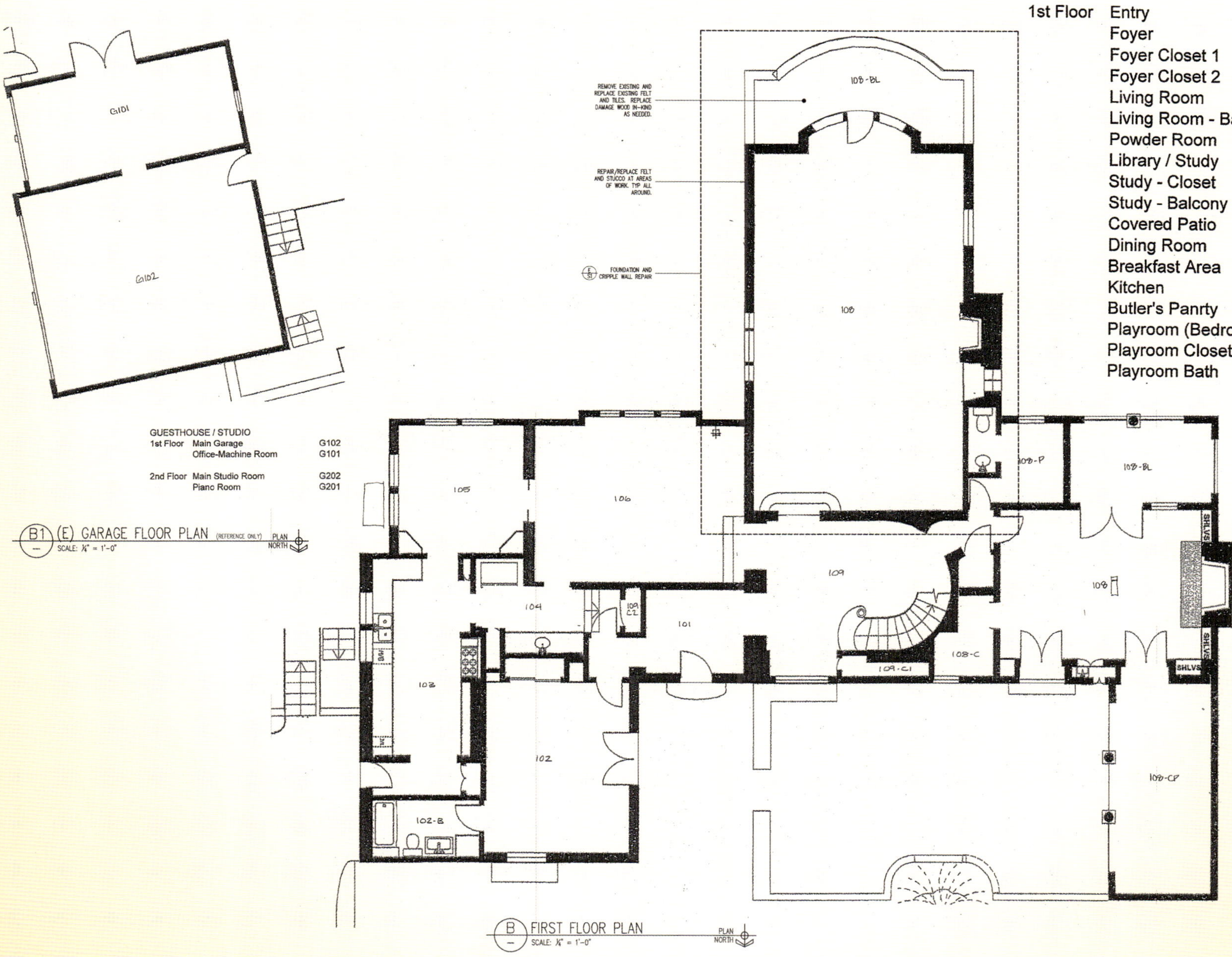

1st Floor		
	Entry	101
	Foyer	109
	Foyer Closet 1	109-C1
	Foyer Closet 2	109-C2
	Living Room	107
	Living Room - Balcony	107-BL
	Powder Room	108-P
	Library / Study	108
	Study - Closet	108-C
	Study - Balcony	108-BL
	Covered Patio	108-CP
	Dining Room	106
	Breakfast Area	105
	Kitchen	103
	Butler's Panrty	104
	Playroom (Bedroom)	102
	Playroom Closet	102-C
	Playroom Bath	102-B

Casa M

M住宅

ARCHITECT | *Federico Martinez Garcia*
FIRM | *Constructora Malver*
PRESIDENT AND OWNER | *Miguel Angel Arce*
DIRECTOR OF CONSTRUCTION | *Carlos Lira Alvarez*
DIRECTOR OF DEVELOPMENT | *Bernardo Arce Valencia*
PHOTOGRAPHER | *Rigoberto Moreno*
LOCATION | *Los Cabos, Mexico*

Built in the most luxurious development in Los Cabos, Mexico, Casa M is a Mexican-style residence with a vocation for relaxation. The natural beauty that surrounds it – a sea of transparent blues and greens – speaks of the interior warmth of the home with its finishes of Talavera and wood. The specialty of MALVER is building custom homes and in this case, the design adjusted to the needs of a young family that required maximum comfort and spaces for enjoying the unique climate of Los Cabos. Therefore, the terrace and open spaces are as ample as is the beautiful swimming pool that continues its dialogue with the beach.

For the owners of Casa M, a decisive element that led them to build a dream is the extraordinary climate of Los Cabos.

这栋建筑位于极度奢华的墨西哥洛斯卡波斯地区，是一栋十分典型的墨西哥住宅，具有明显的休闲娱乐风格。住宅周边是风景秀丽的自然海滨风光，迷人的大海呈现着一种晶莹透亮的蓝色和沁人心脾的绿色。住宅的内部装饰温馨祥和，具有一种西班牙特拉维拉风格。按照马维尔设计公司设计建造房屋的惯例，设计师们为这个年轻的家庭选择了最舒适的设计，提供了各种功能空间，使得业主可以享受洛斯卡波斯独一无二的海滨气候。门廊和其他开放空间，包括游泳池都占有足够宽敞的空间，保持着与外面海滩之间的相互协调。

对于住宅的业主来说，这种决定性的设计元素将为他们实现在洛斯卡波斯独特的风光中畅享人生的梦想。

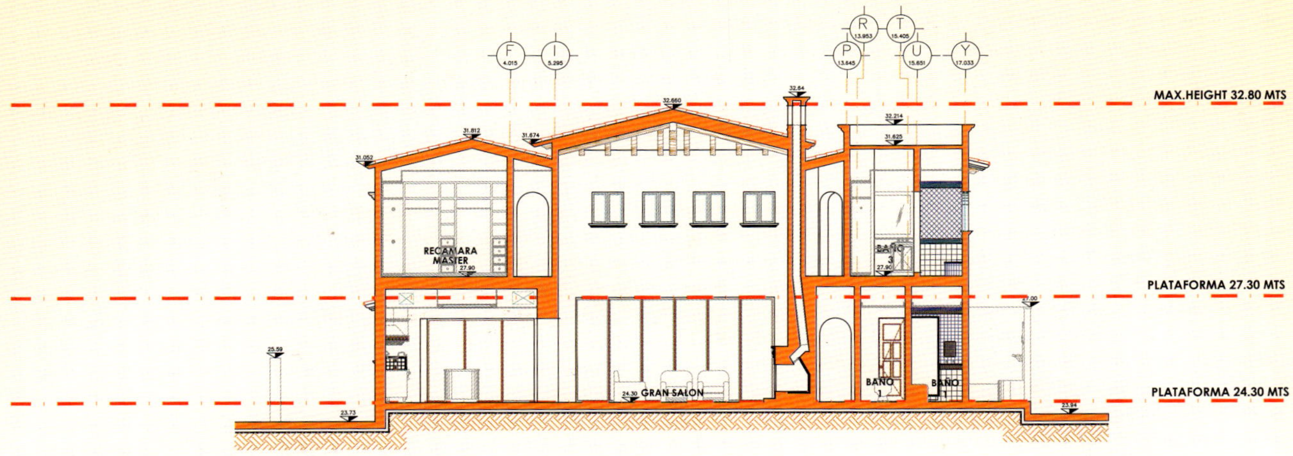

SECCION A—A'

ESC: 1:50

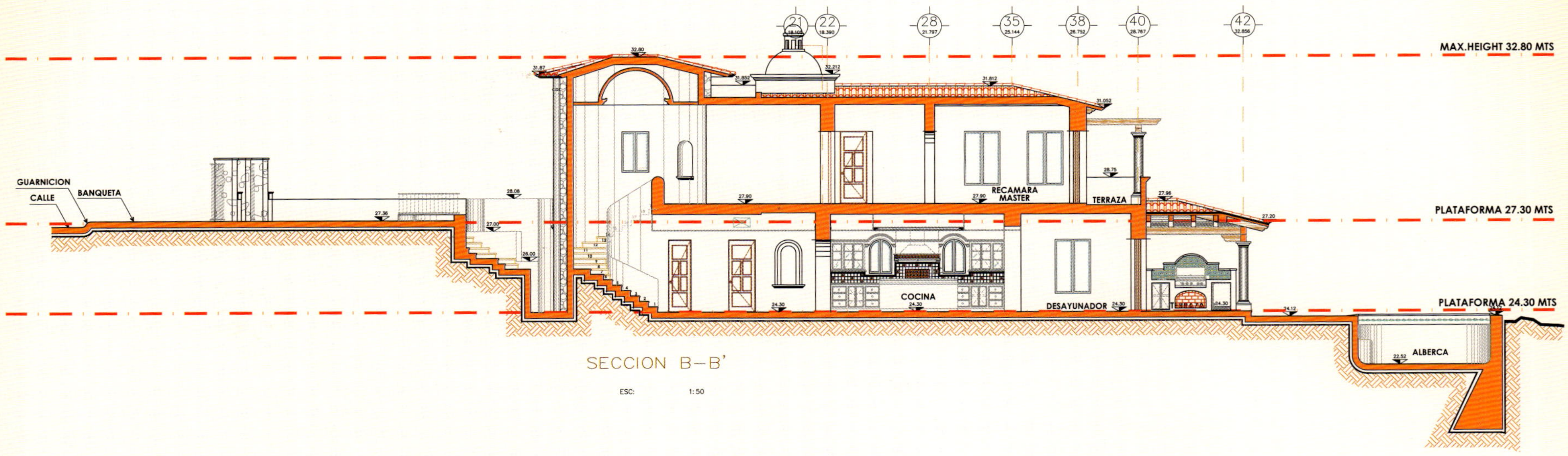

SECCION B—B'

ESC: 1:50

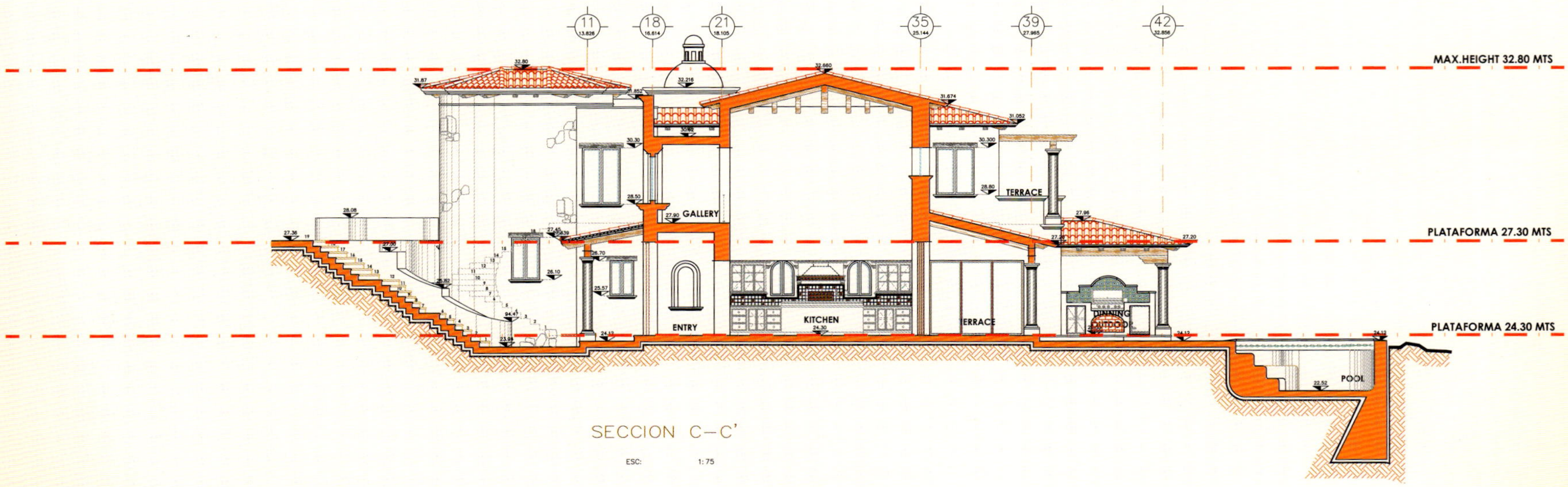

SECCION C—C'

ESC: 1:75

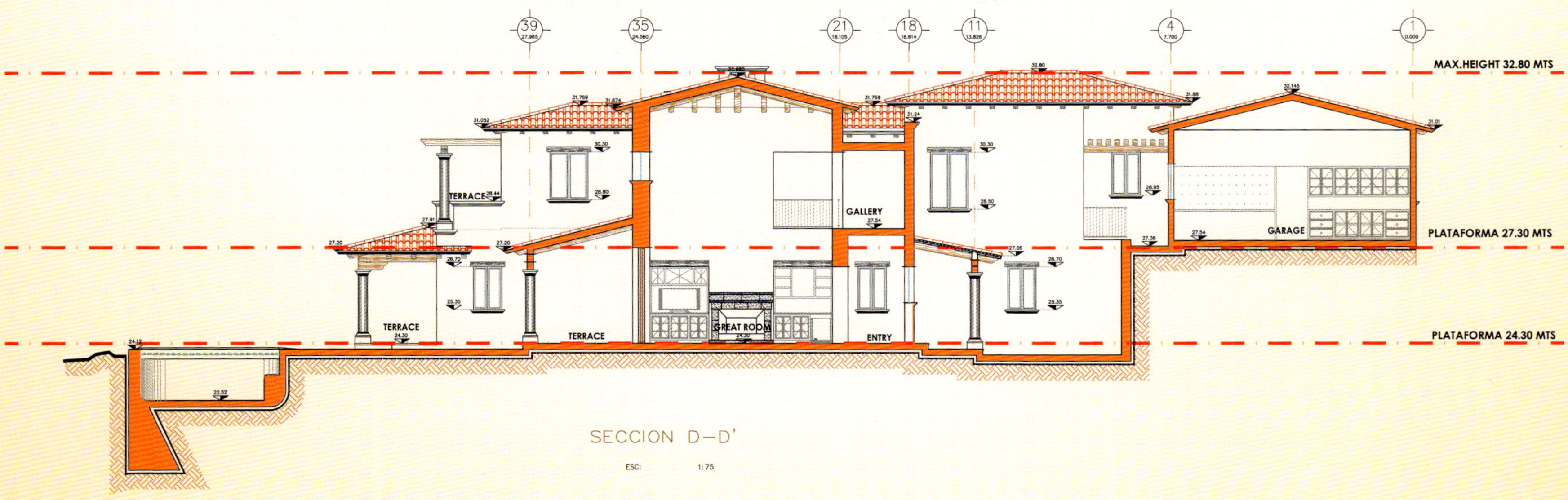

SECCION D—D'

ESC: 1:75

Historic Roland E. Coate

历史悠久的罗兰 · 伊 · 寇特住宅

INTERIOR DESIGNER | Tommy Chambers
DESIGN COMPANY | Tommy Chambers Interiors, Inc.
ARCHITECT | Susan Masterman
LANDSCAPE DESIGNER | Mark Berry
LOCATION | South Pasadena, California, USA
PHOTOGRAPHER | Mieke Kramer and David Phelps

A 1926 Historic Mediterranean estate in South Pasadena, California was designed by Roland E. Coate. The project is a perfect manifestation of Coate's Californian Style with its Monterey Colonial elements.

Meticulously renovated and decorated with a spontaneous mix of traditional and contemporary art and furnishings.

历史悠久的罗兰·伊·寇特住宅坐落在美国加州帕萨迪纳南部，是一处建造于1926年，有着悠久历史的地中海风格建筑。该建筑以其设计师的名字命名，通过蒙特雷殖民元素完美地诠释了寇特式的加利福尼亚风格。

本项目非常注重细节上的修复和建筑整体的粉刷技术，充分体现了传统与现代风格在艺术与装饰方面的完美结合。

Park Lane Residence

公园小巷住宅

DESIGNER | Lloyd Lumpkins, Jeff Salmon, Stacy Brotemarcle, Anne Schooler Kellogg
DESIGN COMPANY | L.Lumpkins Architect, Inc.
BUILDER | Bella Custom Homes
LOCATION | Preston Hollow neighborhood of Dallas, Texas, USA
AREA | 1,394 m²
PHOTOGRAPHER | Lloyd Lumpkins

This residence of over 1,394 square meters is located in the Preston Hollow neighborhood of Dallas, Texas, and designed by Lloyd Lumpkins, a Dallas architect, who is known for designing elegant prestigious homes. The residence completed in 2011, has both formal and informal living spaces, six bedrooms, seven baths, three half baths, four bars, wine cellar and wine tasting rooms, exercise room, exterior covered living area, loggias, balcony, two utility rooms, elevator, four oversized car garages, and a double porte cochere.

This Italian Villa inspired residence is situated on a 4,047 square meters walled compound and landscaped to maximize the use of pleasurable views from each room. As you walk from the entry loggia you are greeted in a large formal marbled reception hall featuring a freestanding curved staircase with custom railings. Just to the left is the formal dining featuring Venetian plaster walls and custom stone fireplace mantel. Connected to the foyer is a long, groined vaulted gallery that serves as a transitional space to the formal living, library, and the master bedroom.

When you enter the library off of the gallery, you find yourself in a secluded space that features a 3.66 meters coffered ceiling and view of the front yard. On axis with the library is a powder bath for the use of the guests, and at the end are two French doors that lead to the master bedroom. The bedroom has a 4.27 meters high arched ceiling with large windows that maximize views of the pool and backyard landscape. In the center of the bedroom is a custom stone fireplace where the owners could share intimate evenings.

A gracious residence designed for entertaining, is truly an Italian villa which also provides intimate spaces for family gatherings.

住宅占地 1394 平方米，位于普雷斯顿霍洛，毗邻德克萨斯的达拉斯，由达拉斯住宅设计领域著名的建筑设计师劳埃德·禄普金斯设计完成。住宅于 2011 年建设完工，里面包括各种生活空间：6 个卧室、7 个浴室、3 个半卫（只有洗手盆和马桶）、4 个吧台，另外还有酒窖和品酒屋、健身房、户外有带有顶棚的活动区、凉廊、楼厅、2 个多功能室、电梯、4 个超大的停车位以及一条双向车道。

这栋意大利式别墅最引人注目之处在于最大限度地利用了周边占地约 4047 平方米、较为复杂的地理环境，使得每一个房间都能欣赏到不同的景观。人们可以沿着凉廊走进大门，这里是一个完全用大理石装饰的接待大厅，里面有一条经典风格的楼梯通向楼上。左侧是餐厅，内有威尼斯风格的粉饰墙体和古典风格的石砌壁炉台。前厅连接一条长长的带有穹顶的拱形长廊，通向起居室、图书室和主卧室。

沿着长廊走进图书室，这里安谧宁静，层高约 3.66 米，上面是格子屋顶，可以俯瞰外面的庭院。图书室后面是供客人使用的浴室，最后是两扇法式大门，通往主卧室。卧室里设有高约 4.27 米的穹顶，巨大的玻璃窗，视野开阔，池塘和后院的美丽景色尽收眼底。卧室中间有一个传统的石制壁炉，主人可以在这里安然入睡。

这是一栋真正意义上的意大利式别墅，住宅的建筑风格雅致，亲切感十足，充分考虑了家人的娱乐休闲需要，并给家人聚会提供了足够的私密空间。

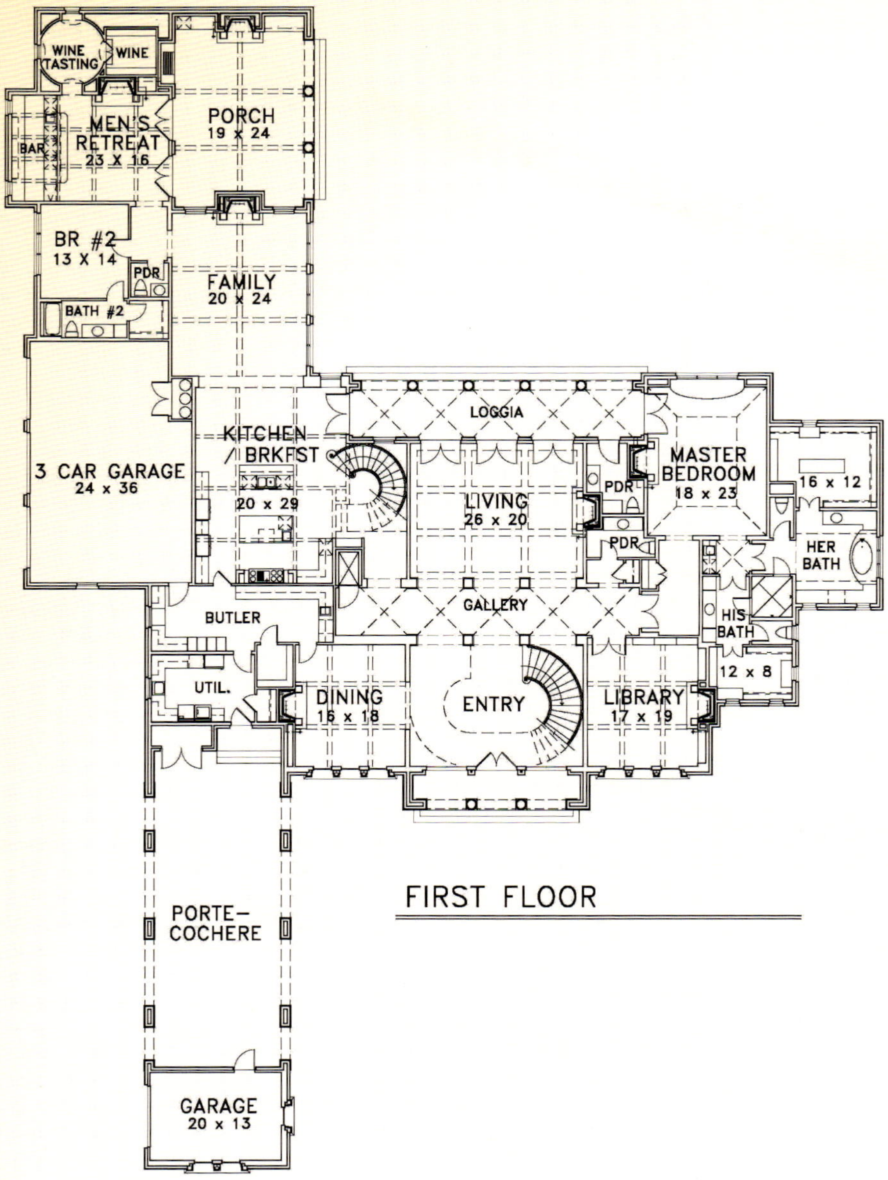

FIRST FLOOR

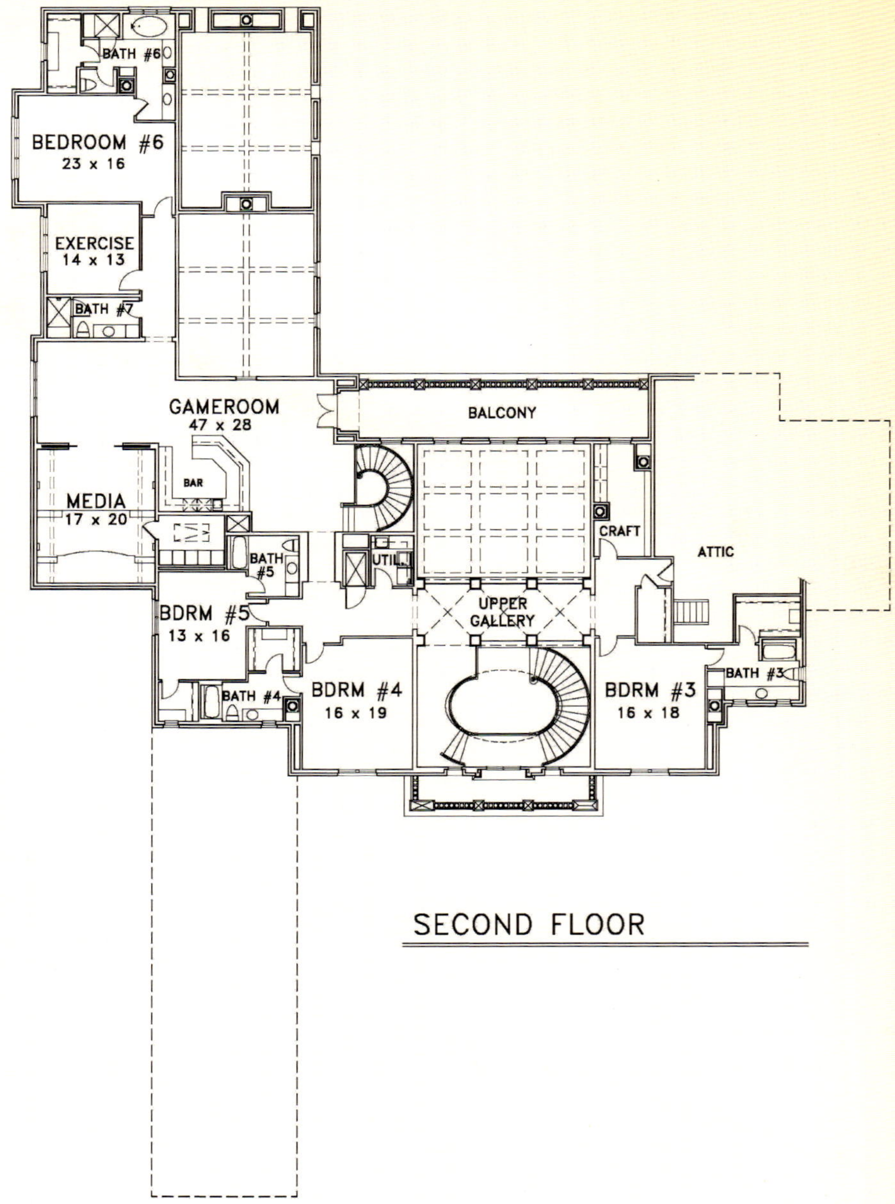

SECOND FLOOR

Casa de la Fuentes

丰特斯住宅

ARCHITECT | *Peikert Group Architects, LLP*

LOCATION | *Los Angeles, USA*

This apartment consists of 42 studios and one bedroom on less than 4,047 square meters of land. The units are targeted for low-income residents who work in downtown Santa Barbara. It is designed in a courtyard housing style reminiscent of the charming Mediterranean style courtyard apartments built circa from 1920s to 1930s in Los Angeles.

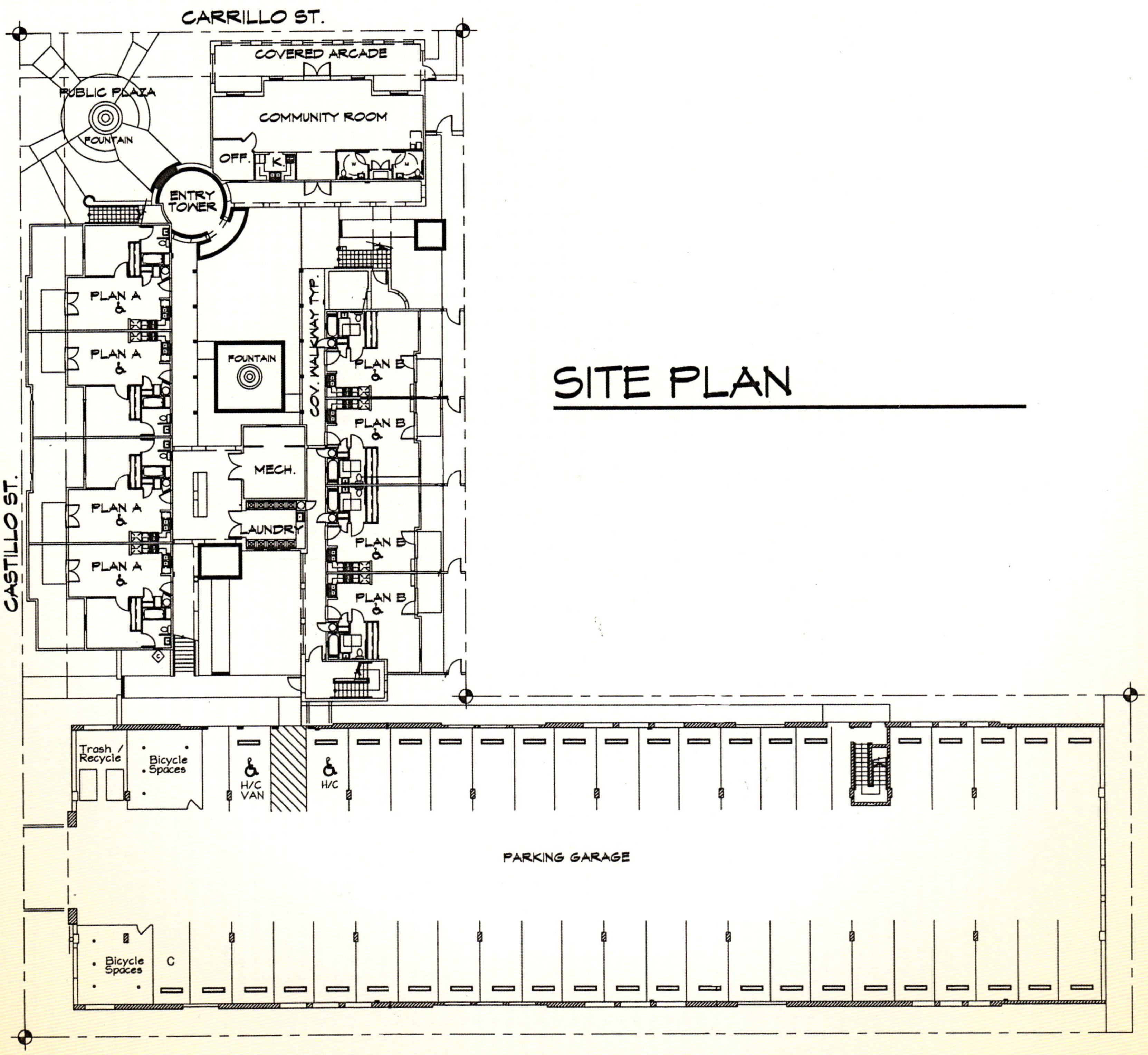

CARRILLO ST.
COVERED ARCADE
PUBLIC PLAZA
FOUNTAIN
COMMUNITY ROOM
OFF.
K.
ENTRY TOWER
PLAN A
PLAN A
FOUNTAIN
COV. WALKWAY TYP.
PLAN B
PLAN B
MECH.
LAUNDRY
PLAN A
PLAN A
PLAN B
PLAN B
CASTILLO ST.
Trash / Recycle
Bicycle Spaces
H/C VAN
H/C
PARKING GARAGE
Bicycle Spaces
C

SITE PLAN

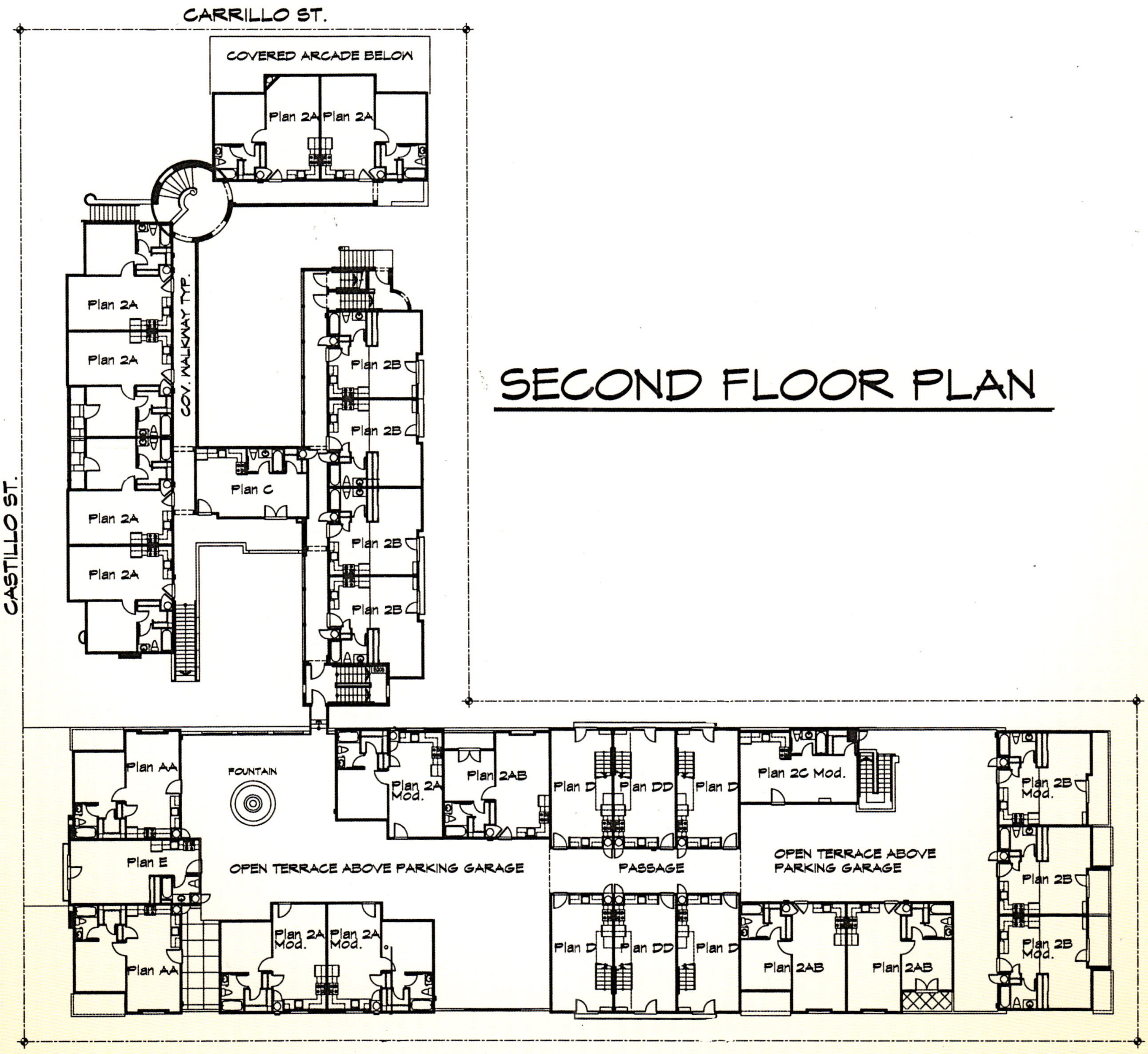

SECOND FLOOR PLAN

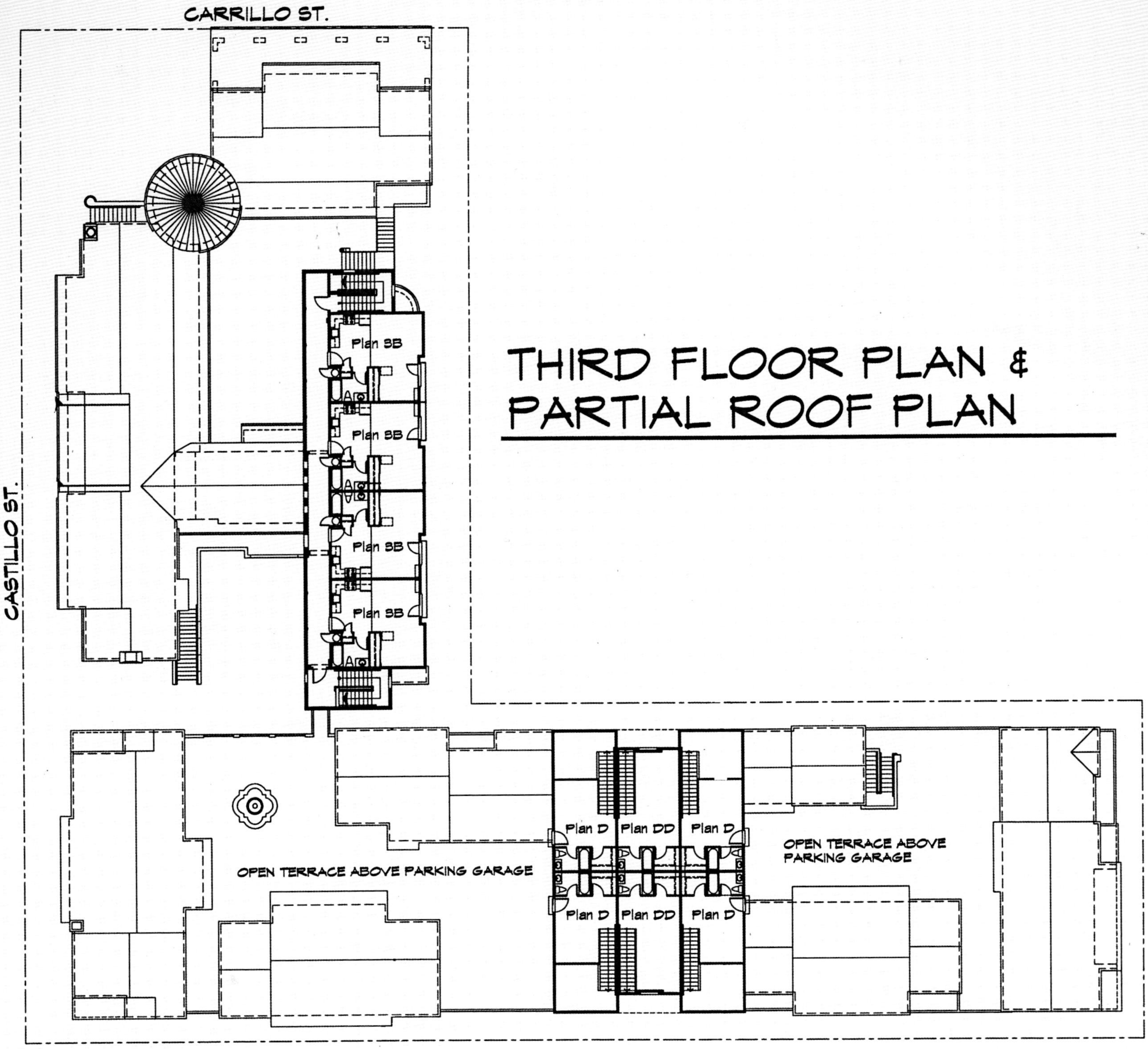

丰特斯房屋是一栋公寓式建筑，占地不到 4047 平方米，包括 42 个工作室和一个卧室。该建筑是专门为那些在圣巴巴拉市中心工作的低收入者设计建造的。设计师们将该公寓设计成具有怀旧感的庭院式建筑，主要参考了 20 世纪二三十年代洛杉矶地区迷人的地中海风格的庭院式公寓。

EXIT

Greenlee Residence

格林利公馆

ARCHITECT | Hugh Jefferson Randolph Architects
PHOTOGRAPHER | Paul Bardagjy

A long triangular lot in West Austin and a client's desire for classic European aesthetic were the driving forces for this design. Inspired by the buildings on triangular lots, the designer had seen while working in Washington, DC, the corner element of the house is round, balancing the simplicity of the rectangle of the main part of the house. Other important goals were simplicity and understatement, ample natural light, and breaking down of the exterior scale to fit into an existing neighborhood. Interior design was done in collaboration with Mark Ashby Designs.

Main Floor

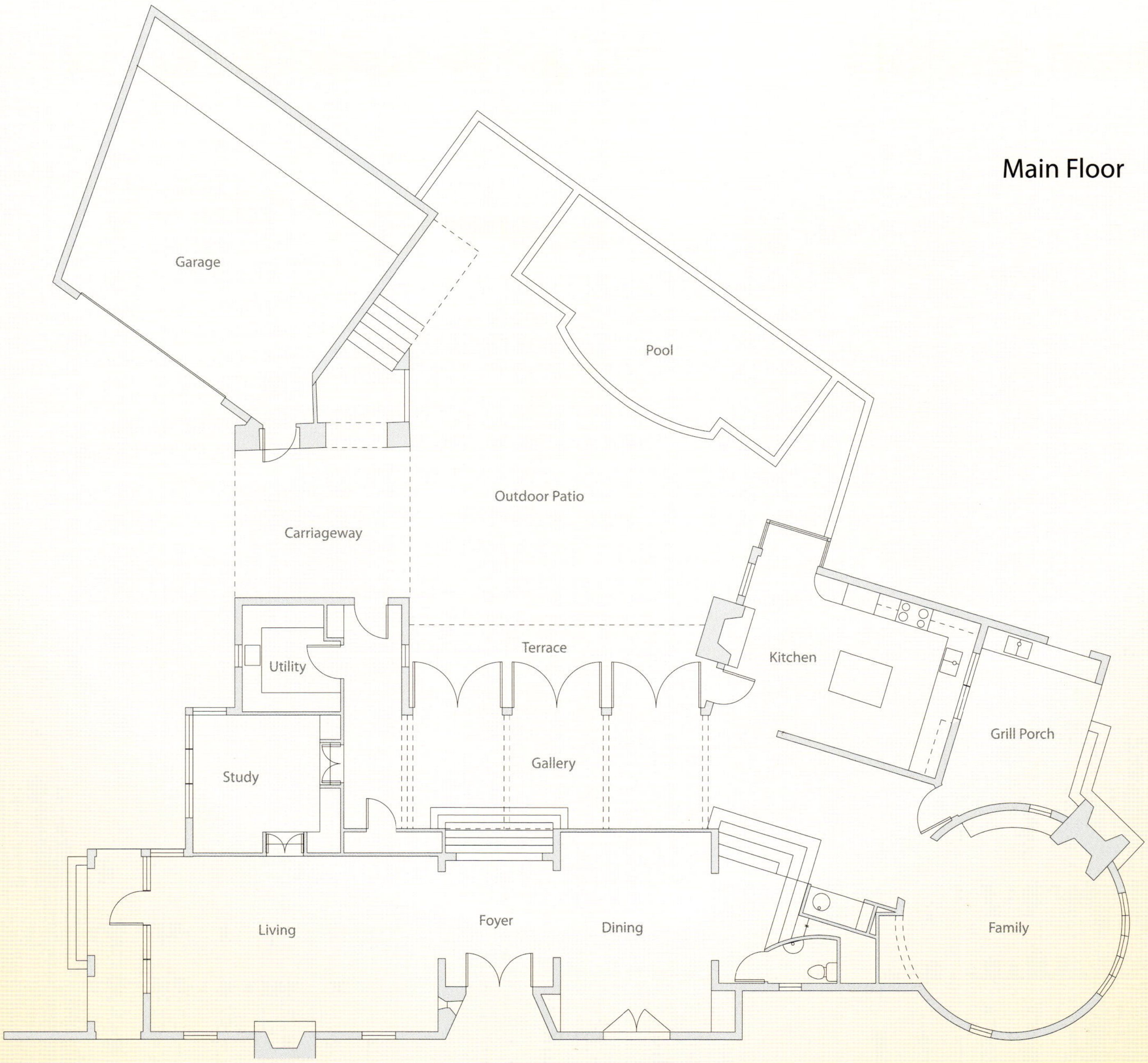

Garage
Pool
Outdoor Patio
Carriageway
Terrace
Kitchen
Utility
Gallery
Grill Porch
Study
Living
Foyer
Dining
Family

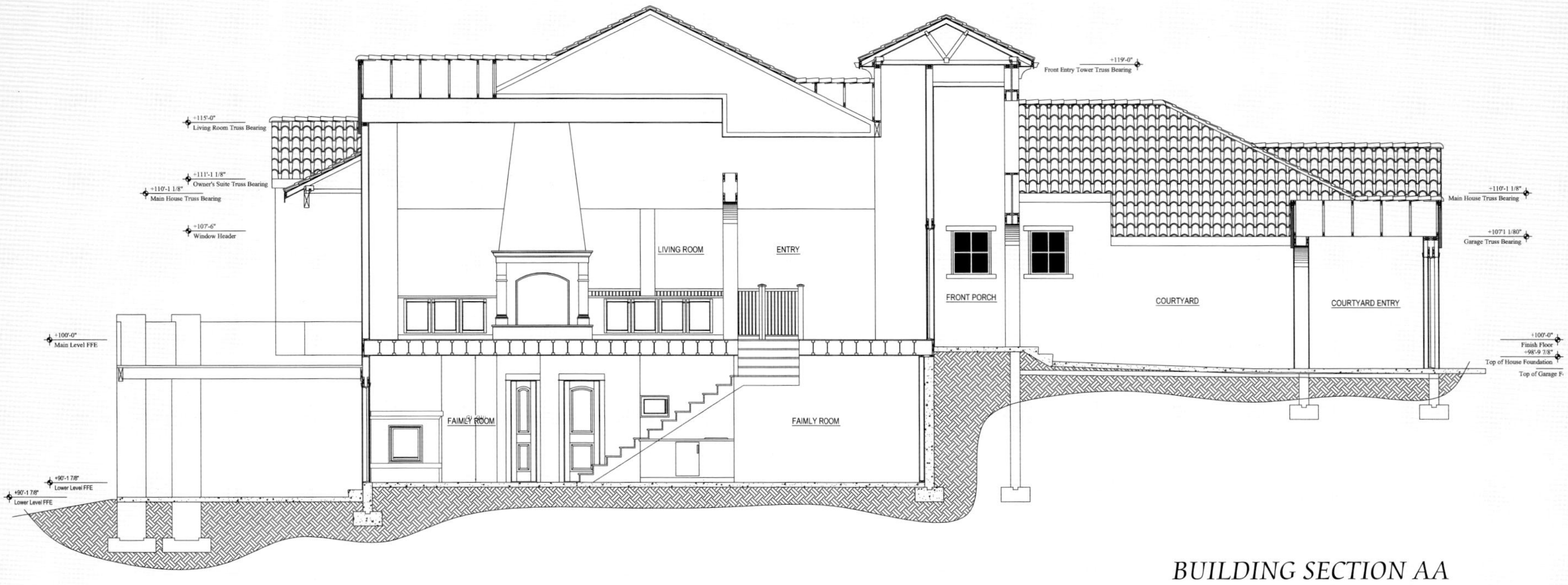

BUILDING SECTION AA

BUILDING SECTION BB

1 FRONT ELEVATION

这个住宅的家居设计温馨舒适，融合了优雅和成熟的气息。起居室户外是舒适的庭院，旁边设有一个石壁炉，这里是感受凉爽的科罗拉多夜晚的最佳去处。宽敞的客厅也设有一个由人造石建造的豪华壁炉，两侧分立着高大的书架。地板选用宽而厚的桃木，给整个住宅增添了温馨的感觉。餐厅和厨房的上方是枝型的水晶吊灯，主人浴室里设有丝滑柔顺的褶皱布帘，这些元素无不给人以动人心魄的优雅之感。住宅内还单独设立了盐浴房间，具有典型的古代亚洲阁楼的设计风格，墙上贴满了印花纹壁纸，隐约透露出一丝淡淡的浮华。

1 FLOORPLAN LOWER LEVEL

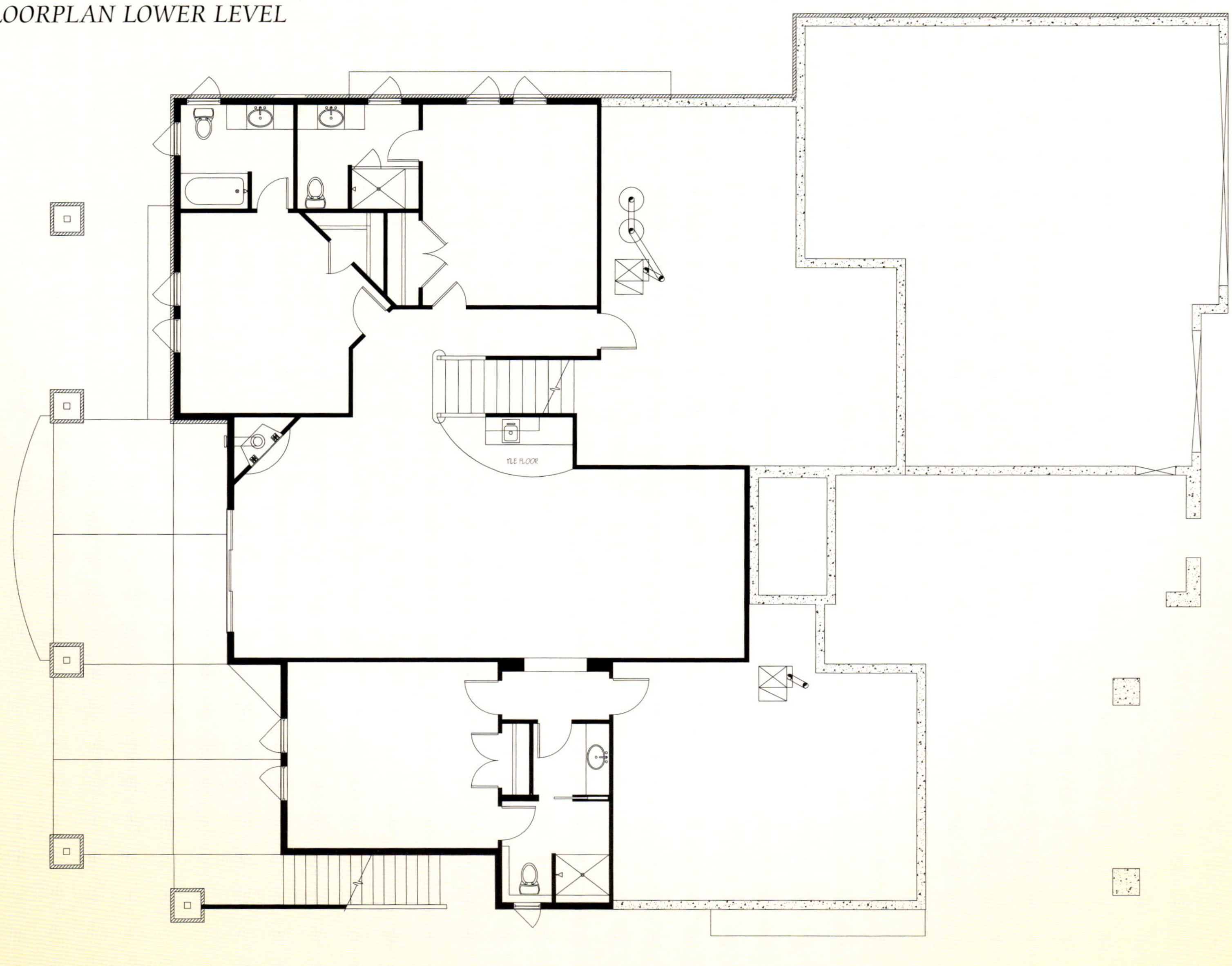

1 FLOORPLAN MAIN LEVEL

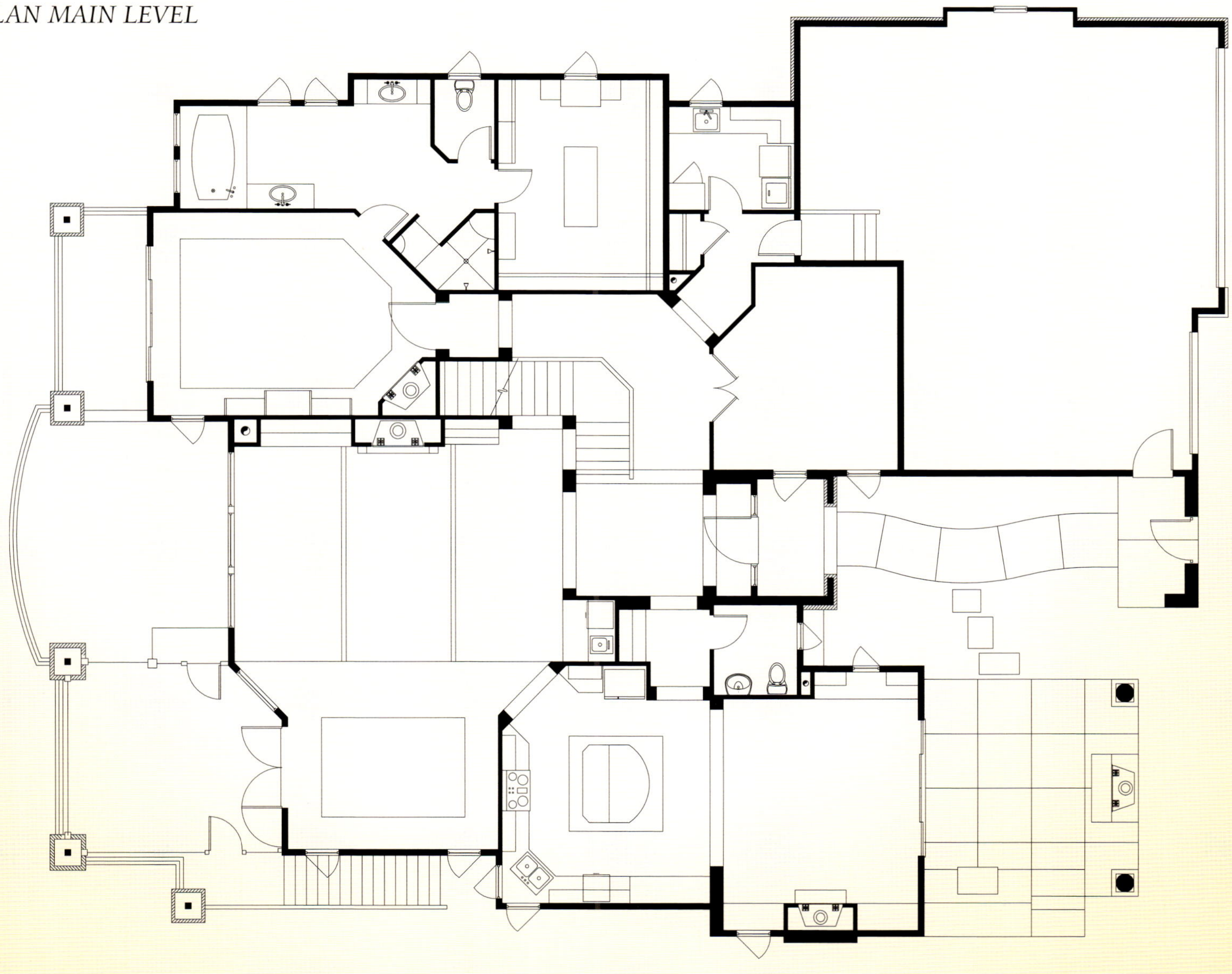

Mediterranean Style Residence

地中海式住宅

DESIGNER | Tom Taylor
DESIGN COMPANY | Taylor Lombardo Architects
LOCATION | Portola Valley, California, USA
AREA | 883 m²
PHOTOGRAPHER | Adrian Gregorutti

With nearly 883 square meters of main house, guest house, and pool house, the project also has several inviting outdoor-living areas to enjoy the beautifully landscaped 10,522 square meters site. This includes four outdoor fireplaces, each displaying a different type of antique Italian firebrick. Several reclaimed materials were used throughout the residence to give it an authentic old world look.

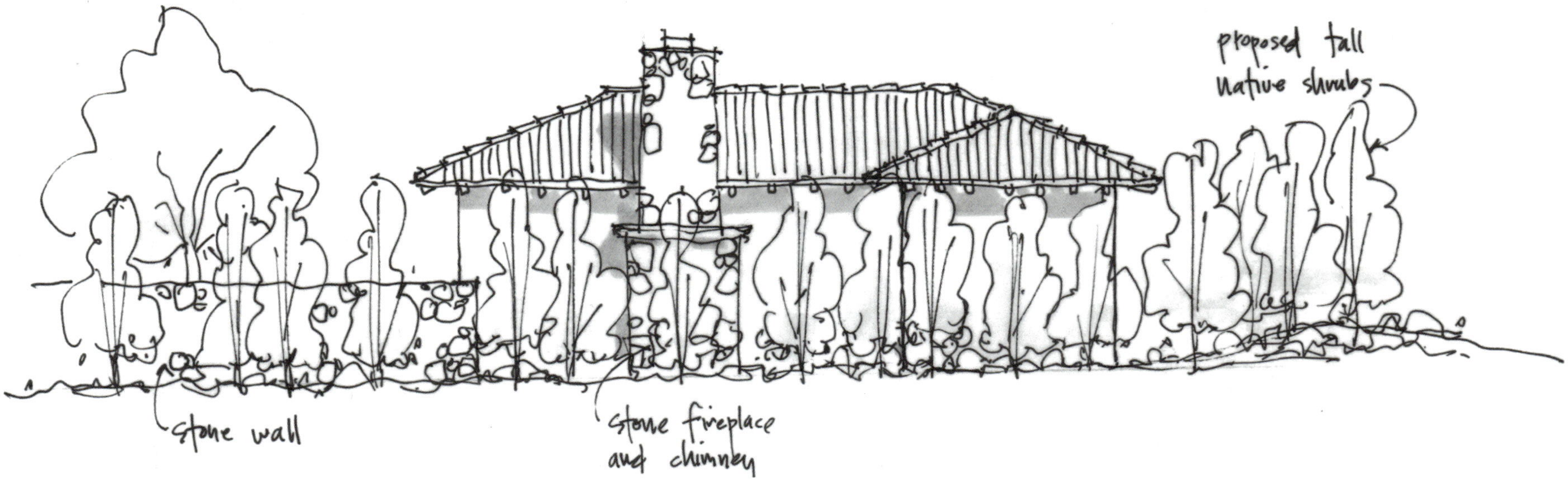
proposed tall native shrubs
stone wall
stone fireplace and chimney

别墅的建筑主体、客房以及游泳池共占地约883平方米。此外，该项目还拥有独特的露天生活区，占地10522平方米，景色宜人，以供人们娱乐消遣。这里设计有四个室外壁炉，每一个都采用不同类型的意大利仿古耐火砖砌成，展现出与众不同的风格。整个别墅均采用独特的、可再生的建筑材料，远远望去，颇具几分古代建筑的神韵。

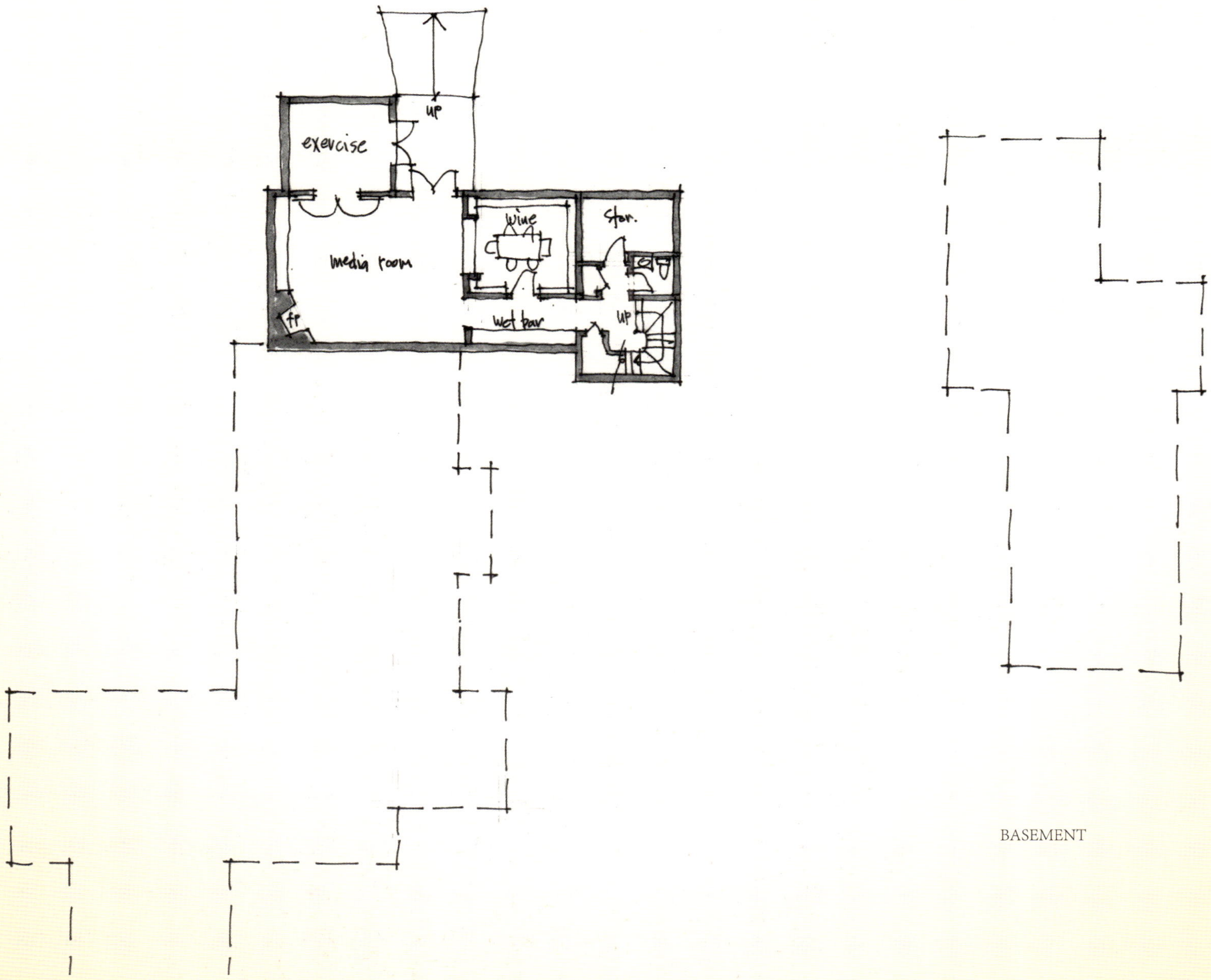

BASEMENT

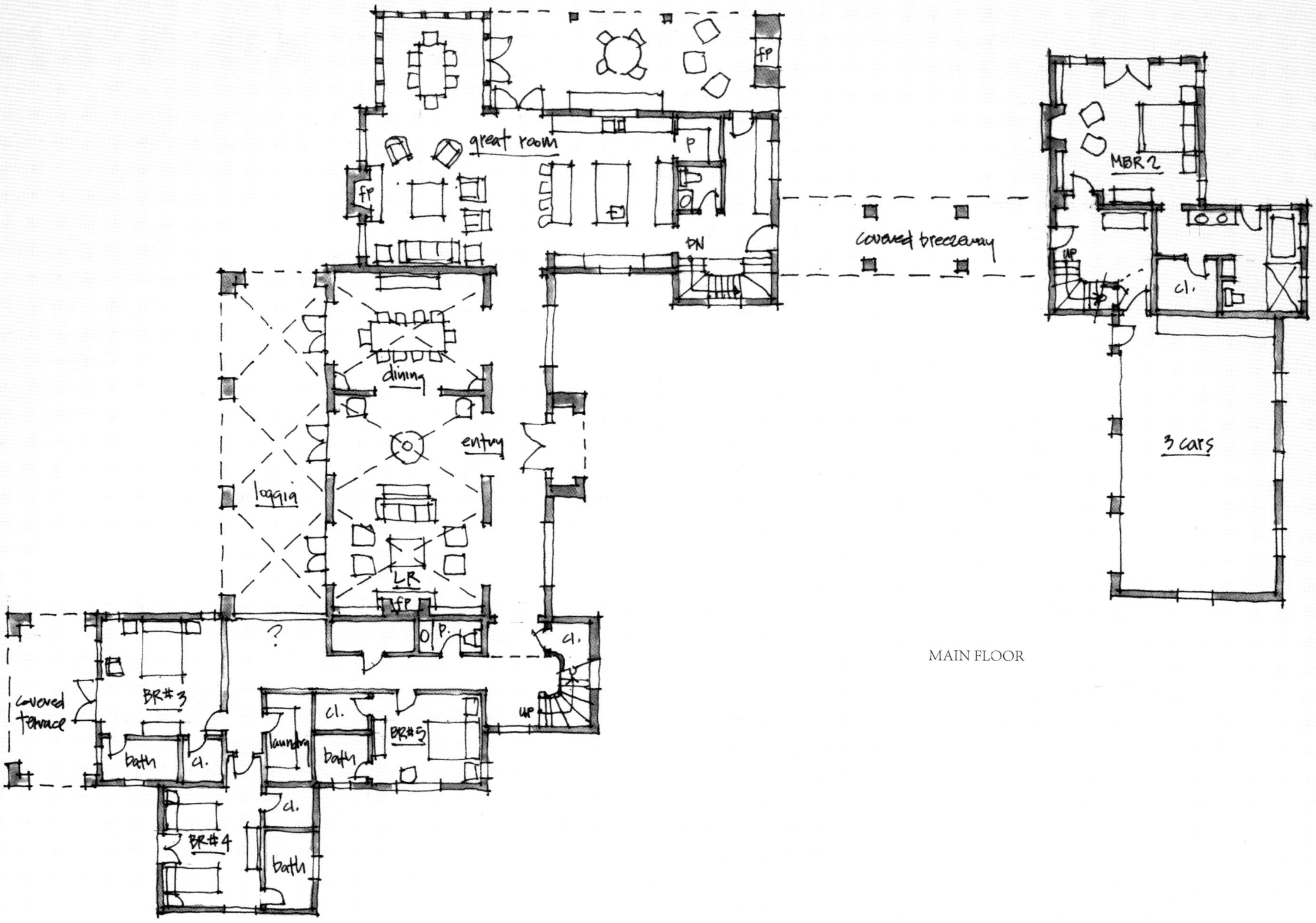

MAIN FLOOR

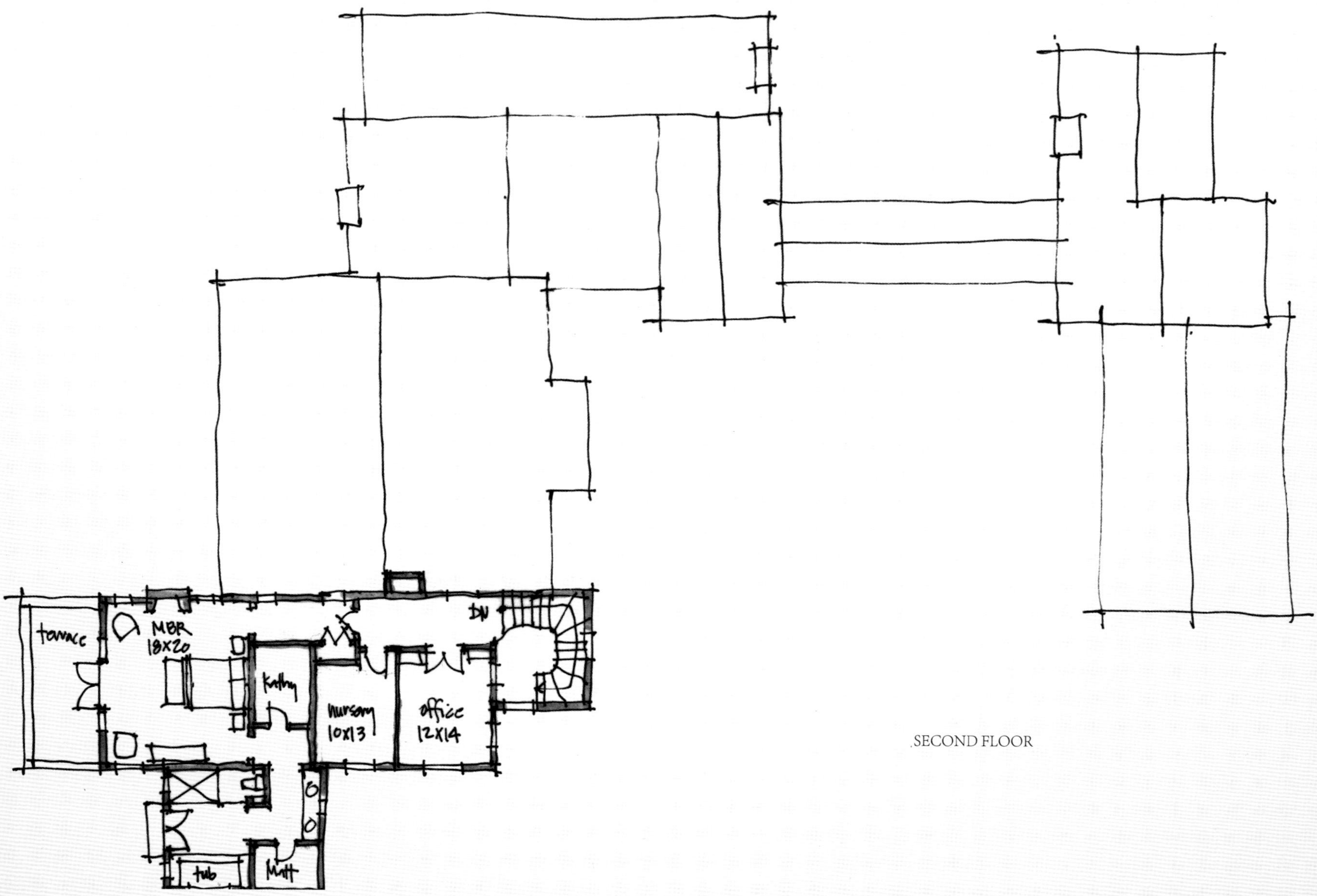

SECOND FLOOR

INDEX

索 引

Jackie Armour

Founder and President of JMA Interior Decoration, Inc. is an Allied Member of the American Society of Interior Designers. She has an AS in Interior Design with concentrations in architecture as well as a BS in Business Administration from the University of Florida. With 10 years of experience, Jackie and her team work closely with each client to create classic, timeless interiors. Detailed floor plans, architectural elevations, and interior perspectives create the foundation of the design process with JMA. Custom designed furniture, fabrics, window treatments, floor coverings, bedding, color and lighting uniquely define a comprehensive project.

Deborah Costa

Deborah Costa is President of Deborah Costa Design Alchemy, one of the leading design services and home merchandising firms in Northern California. Focused on understanding the needs, dreams and desires of their clients and creating spaces that exceed expectations the firm has built a solid reputation through creating, transforming and beautifying projects across all spectrums and geographies.

Inspired by old world architecture, new technologies and enhancing people's lifestyles, the firm believes in the uniqueness of each design ensuring that the result evokes the personality of their clients. Collaborative in nature, Deborah and her talented team appreciate that life is fluid so design and process must constantly evolve to maintain significance. The firm is driven by the constant innovation of ideas, style and implementation while building long-standing relationships with clients and artisans around the globe.

Carson Poetzl Inc.

They are a full-service architectural firm that has a rich history and focus dedicated to high-end custom residential and light commercial architecture. They approach the highest level of design and service by closely working with builders, interior designers, landscape architects and engineers. They offer a wide range of talents that are represented in designing styles ranging from cutting-edge contemporary to historically accurate. They tailor each project around the clients' needs and desires, while integrating climatic and site-specific ideas to create a blended harmony between the architecture and its surroundings. They are an award winning firm that has appeared in numerous publications.

The Hotel Hershey

The Hotel Hershey is one of America's landmark hotels, offering 276 guest rooms and 2,323 square meters of meeting and function space. The Hotel Hershey is part of Hershey Resorts, which also includes Hershey Lodge and Hershey Highmeadow Campground. Hershey is conveniently located three hours from New York City, two hours from Philadelphia and Washington, 90 minutes from Baltimore and four hours from Pittsburgh. Harrisburg International Airport (MDT) is just 20 minutes from Hershey.

Tommy Chambers

Los Angeles interior designer Tommy Chambers opened his own design company, Tommy Chambers Interiors Incorporated, in 2001 after eight years at Chambers and Murray, Inc. as head of the interiors department and the senior business and operations manager. Prior to starting Chambers and Murray, Inc. Tommy apprenticed for five years with the venerable interior designer, Joan Axelrod of Joan Axelrod Interiors.

In addition to receiving an architecture degree from Texas A&M University, Tommy spent a year in Europe studying art, architecture and decorative arts.

Tommy is listed in the Los Angeles Franklin Report, an independent evaluator of designers and design services. He is also an active professional member of the American Society of Interior Designers (ASID), the International Interior Design Association (IIDA) and the Institute of Classical Art and Architecture (ICAA).

Vanguard Studio Inc.

Vanguard Studio Inc. is an Architectural Design Firm specializing in custom and speculative homes in the Central Texas area. Their award winning designs strive to combine the timeless qualities of the past and progressive ideas of today. They have developed an innovative and efficient design process, which enables them to achieve superior results and exceed clients' expectations. Vanguard has successfully designed hundreds of projects since its creation in January of 2000, with budgets ranging from $300k to several millions. They believe that strong design which is responsive to the client's needs and dreams is always possible – no matter what constraints exist. They work to assemble a design team, which also includes a builder, interior designer, and other various consultants, to ensure that their client's vision is fully explored while meeting their programmatic and budgetary needs. Vanguard has talented staff that it employs to accomplish these goals.

Taylor Lombardo Architects

Taylor Lombardo Architects is an award winning Architecture and Land Planning firm with offices in San Francisco and Napa Valley, California. In addition to a large number of projects in California, the firm has completed projects throughout the United States, and internationally in China. The firm specializes in custom homes and wineries of all sizes and styles, tailored specifically to the needs of each client. Taylor Lombardo Architects have designed over 30 wineries. The firm also has extensive experience in hotel, retail, and restaurant design.

Partners, Tom Taylor and Maurice Lombardo are the primary client contact and remain very hands-on throughout the project. Taylor Lombardo Architects' philosophy maintains that the firm works closely with the client, ensuring that each project is a creation specific to the client's desires. The firm's design approach is one of creating site sensitive and environmentally responsible structures that are both timeless, innovative, and enduring. Careful attention to architectural scale, proportion, detailing, and material selection assures the design will be true to the style desired.

Columbia Hospitality, Inc.

Columbia Hospitality, Inc. is a leading hospitality management and consulting firm specializing in luxury hotel and resort management. Columbia Hospitality's premier properties include boutique hotels and secluded resorts.

www.columbiahospitality.com

Adele Chang, AIA

Adele Chang received her bachelor's degree in architecture from the California State Polytechnic University at Pomona in 1980 and is a licensed architect in the State of California. Her extensive design background is complemented by expertise in site planning and experience in working with various municipal agencies and neighborhood groups.

Adele Chang is a member of the American Institute of Architects, has served as a judge for various national and regional design awards, has been featured as a speaker on housing design at numerous building industry conferences and workshops. The work of Adele Chang and Lim Chang Rohling Associates is recognized for its excellence with numerous awards and in various publications.

Fabrizia Frezza

Fabrizia Frezza, Italian-born, got a first-class degree in Architecture from the University of "La Sapienza" in Rome, Italy. She has been on the Register of Architects since 1990. Her love of art, in all forms, her travels around the world, her encounters with different cultures have made an impression on her creative spirit, enriching her knowledge and curiosity.

Fabrizia Frezza was born and raised in family hotels and understands the needs of this sector. The architect uses her expertise and works with ease when planning and renovating hotels. She unites the rationalization of space, incorporating domestic needs, with hotel comforts making the hotels a home away from home. The furnishings, fabrics and materials are all selectively chosen for each ambiance. Her skills, high standards and continuous desire to go beyond brings her to work alongside artisans and up-and-coming artists uniting their work, thoughts and creativity.

Peikert Group Architects, LLP

Peikert Group Architects, LLP is a recognized leader in the design of high quality market-rate and affordable multi-family housing, senior housing, mixed use, and commercial projects along the South Coast of California. The firm has received over sixty local, state and national design awards.

House + House Architects

For 30 years House + House Architects have produced a diverse body of work reflecting their passion for site specific, well choreographed buildings throughout California, Mexico and the Caribbean. Committed to sustainability, their work incorporates natural light, ventilation, passive and solar heating, grey/rain water systems and recycled materials.

House + House Architects have received over 50 design awards and their work has been featured in numerous national and international publications including two monographs, "House + House Architects: Choreographing Space" and "Houses in the Sun: light movement embrace". Their stunning book, "Mediterranean Villages: An Architectural Journey", celebrates their investigations of vernacular architecture throughout the Mediterranean.

Steven and Cathi House have lectured extensively throughout the United States and Mexico and have both served on the Advisory Council for Virginia Tech's College of Architecture. They have recently established CASA a new architecture study abroad program based in Mexico, with the goal to instill in their students a holistic view of what it means to be an architect in a globally interdependent and culturally diverse world.

Chris Barrett

Chris Barrett Design is inspired at its roots by the brilliant light and clean sensibility of the Californian landscape.

Founded in 1987 and named for its owner – interior designer and Southern California native Chris Barrett – the firm is renowned for its breathtaking residential interiors and chic commercial spaces. Alchemists with color, light, pattern and scale, Barrett and her team execute bold visions for their fast-growing portfolio of clients, consistently revealing functionally creative living spaces with a balanced look that is both dramatic yet restrained.

In keeping with the designer's profound love of the land, Barrett also develops her design projects in a way that encourages the principles of environmental sustainability. Livable outdoor spaces that extend naturally from fresh, light-filled interiors are a hallmark of the designers' work, as are interiors that reflect the spaciousness and order of nature.

Constructora Malver

Constructora Malver defines itself as a company that offers services of excellence to clients that are accustomed to the very best. It is only through this excellence that clients can live the building process as a gratifying experience, and truly enjoy their homes.

The excellence in services implies high quality standards in the construction, which is the only way to assure a satisfactory delivery and an efficient functioning of the product.

In its dynamic construction work, Malver capitalizes on cumulative experience and has been enriched by a human team that shares and protects the principles that sustain the company: order, honesty, professionalism, and legality. As a certified company, it meets all its legal duties and obligations.

Rentfrow Design, LLC

Rentfrow Design, LLC. was founded by Jon Rentfrow in 1995. Jon established the business to follow his passion for architecture. He has strived to develop a company that glorifies his Lord and Savior, Jesus Christ, by focusing on maintaining long-term relationships with clients. Jon believes building a relationship begins with satisfying the customer with a design that meets their unique desires and personal lifestyles. Jon's approach is to design a distinctive home by "Creating Moods in Individual Spaces". Jon is an award-winning designer and is a Professional Member of the American Institute of Building Designers (AIBD). Jon received his Bachelor of Science degree in Interior Design from Colorado State University in 1994. Jon works out of his personal residence to afford time with his wife, Elisha and three sons, Nicholas, Jathan, and Zekiah. Jon enjoys riding his bike, playing softball and golf, and investing time at his church.

Design Vidal

Karen and Guy Vidal, the creative forces behind the Design Vidal Interior Design team, have provided interior and exterior design services to over 100 residences in the Los Angeles area. Karen Vidal's passion for restoring and remodeling homes with historic and classic architecture has enabled her to become the "designer of choice" in Silver Lake, Los Feliz and surrounding areas. Design Vidal transformations have also taken place in Santa Monica, Venice, Glendale, Monrovia, Pasadena, Hancock Park and Bel Air.

Praised by their peers for their ability to work in virtually any style of home, the Design Vidal Team can add their design expertise to your home regardless of its architectural style. Known in historic neighborhoods as the "Design Green Team", Karen and Guy Vidal take pride in restoring and recycling many formerly classified "tear-down homes".

Elevation Architecture Studios

Elevation is Southern California's Premiere luxury residential architecture firm. Located in the Del Mar and Rancho Santa Fe area, their experienced team of designers, architects and computer expert are providing their clients timeless, elegant design solutions tailored to fit their lifestyles.

They are a customer-oriented company that prides itself on listening to their clients' needs and then creating their clients' vision. They are team players and know the value of interfacing harmoniously with engineers, interior designers, consultants and builders. Their designers are well versed in today's design styles and trends, green solutions and they use chef consultants for their kitchen designs. California is about indoor/outdoor living and that is what they specialize in designing.

DDG

DDG is an internationally renowned planning, architecture and design company unique in its ability to integrate multiple disciplines and to create memorable places all over the world – Town Square Las Vegas, Pondok Indah Mall, Muvico Theaters, Istinye Park. The firm's successful track record is reinforced over time and again with numerous awards received for designs that work in the real world. DDG will formulate the highest and best use for property, utilizing their strongest asset: creative design.

Robert Hidey

As one of the world's most renowned residential architects, Robert Hidey, AIA, has pioneered housing trends for traditional luxury homes in settings as diverse as Irvine and India, Beverly Hills and Beijing, Durango (Colorado) and Dubai.

Since its establishment nearly two decades ago, Robert Hidey Architects (RHA) has initiated some of the most original and creative floor plan designs and land planning in the industry. By taking small details, such as the precise curve of a roofline or arch, and imbuing them with a sense of authenticity, RHA's diverse projects around the world evoke a common respect for art and history, while creatively accommodating the needs of today.

Aneka Interiors Inc.

Aneka Interiors Inc. is a full service interior design firm specializing in residential & commercial interior architecture & design. Aneka graduated with a Bachelor of Science in Interior Design in 1996 and has been working in the field of design for 18 years. They strive to make each home, building or space they design as unique as the people it serves. Their interior design team specializes in integrating the elements of a design to make it flow consistently thru the space. They select each individual item to work in the big picture, while always thinking of the end result.

Hugh Jefferson Randolph Architects

Hugh Jefferson Randolph Architects is a design focused practice with an emphasis on residences that are specific to their site. Based in Austin, Texas, with an international practice, they approach each project as a collaboration between the architect and the client. Starting with the goals and dreams of the client, they like to structure the early design meetings as brainstorming sessions, where everyone is given a voice, and all suggestions are given, considered and explored.

Even in situations where the design preconceptions are strong, they feel that the most appropriate and freshest solutions are typically not what the architects or the clients expected at the beginning of the process. For them, the best compliment to give or receive is "I would never have thought of that on my own".

Whether the home is a modern and sleek, or more classic and traditional, they strive for each design to be a true reflection of its place and of those who live there. That design philosophy, paired with the highest quality of craftsmanship, attention to detail and respect for the natural environment has been a constant for them for over twenty years. They are always very honored to be entrusted with the responsibility of helping families create their homes, and know that each and every project is a special one.

Cornerstone Architects

Founded in 1989, Cornerstone Architects is an architectural design firm based in Austin, Texas. Cornerstone specializes in architecture and interior design for both residential and commercial projects. Cornerstone's international experience includes both commercial and luxury residential projects. Their varied portfolio consists of retail, restaurant, recreational, resort, multi-family, office, medical, religious, and single family residential projects. The firm's design approach blends architectural, interior, and sustainable design disciplines to create a uniquely balanced cutting edge architectural solution for each project. Cornerstone Architects has received numerous awards including several National Commercial Builders Council Awards of Excellence. As a participant in many homes tours and parade of homes, Cornerstone has also gained numerous accolades for their residential work.

ACKNOWLEDGEMENTS

We would like to thank everyone involved in the production of this book, especially all the artists, designers, architects and photographers for their kind permission to publish their works. We are also very grateful to many other people whose names do not appear on the credits but who provided assistance and support. We highly appreciate the contribution of images, ideas, and concepts and thank them for allowing their creativity to be shared with readers around the world.

本书的译者为大连工业大学外国语学院的何心。